THIS SECOND EDITION OF *Designing Tall Buildings*, an accessible reference to guide you through the fundamental principles of designing high-rises, features two new chapters, additional sections, 400 images, project examples, and updated US and international codes. Each chapter focuses on a theme central to tall building design, giving a comprehensive overview of the related architecture and structural engineering concepts. Author Mark Sarkisian, PE, SE, LEED® AP BD+C, provides clear definitions of technical terms and introduces important equations, gradually developing your knowledge. Projects drawn from SOM's vast portfolio of built high-rises, many of which Sarkisian engineered, demonstrate these concepts.

This book advises you to consider the influence of a particular site's geology, wind conditions, and seismicity. Using this contextual knowledge and analysis, you can determine what types of structural solutions are best suited for a tower on that site. You can then conceptualize and devise efficient structural systems that are not only safe, but also constructible and economical. Sarkisian also addresses the influence of nature in design, urging you to integrate structure and architecture for buildings of superior performance, sustainability, and aesthetic excellence.

Mark Sarkisian, PE, SE, LEED® AP BD+C, is the Partner-in-Charge of Seismic and Structural Engineering in the San Francisco office of SOM. He has developed innovative structural solutions for more than 100 international building projects, including some of the tallest buildings ever constructed. A world-renowned leader in the design of high performance seismic and environmentally responsible structural systems, Sarkisian has patented numerous inventions and has additional patents pending. He teaches, publishes, and lectures frequently around the world.

"*Designing Tall Buildings* is *the* best book on structures and architecture available. It appeals to both architects and engineers, capturing the "why" and the "how" of tall building construction. It explains the history of tall buildings and details the structural rationale behind the art. Mark Sarkisian is a brilliant and engaging instructor, bringing structural theory to life. This book is a must-read for all designers." – **Mary Comerio, Professor of the Graduate School, UC Berkeley, USA**

"*Designing Tall Buildings* is a seminal reference guide that clearly illustrates the inseparability of architecture, structural design, and local context in the realization of tall buildings around the world. As one of SOM's foremost structural engineering partners and a global thought leader on skyscraper design, Mark Sarkisian is the perfect author to give this message." – **Antony Wood, Executive Director, Council on Tall Buildings and Urban Habitat, USA**

"*Designing Tall Buildings: Structure as Architecture* clearly discusses the roles that structural design and nature play in tall buildings. Mark Sarkisian communicates an intuitive understanding of the interrelationships between forces at play. Clear and well written, with definitions at the end of each chapter, this book serves as an excellent learning tool." – **Jon Daniel Davey, AIA, Professor of Architecture, Southern Illinois University, USA**

"Sarkisian's *Designing Tall Buildings* provides a masterful discussion of the synergy of architecture and structural engineering in landmark building designs by SOM. This book has inspired both students and faculty in design studio courses at Stanford University that aspire to combine technical rigor and creative thinking in architecture and engineering through the art of tall building design." – **Greg Deierlein, Department of Civil and Environmental Engineering, Stanford University, USA**

DESIGNING TALL BUILDINGS

STRUCTURE AS ARCHITECTURE

SECOND EDITION

MARK SARKISIAN

NEW YORK AND LONDON

First published 2016
by Routledge
711 Third Avenue, New York, NY 10017

and by Routledge
2 Park Square, Milton Park, Abingdon, Oxon OX14 4RN

Routledge is an imprint of the Taylor & Francis Group, an informa business

Library of Congress Cataloguing in Publication Data
Sarkisian, Mark P., author.
Designing tall buildings : structure as architecture / Mark Sarkisian. -- Second edition.
pages cm
Includes bibliographical references and index.
1. Tall buildings--Design and construction. I. Title.
NA6230.S27 2016
720'.483--dc23
2015023170

ISBN: 978-1-138-88670-4 (hbk)
ISBN: 978-1-138-88671-1 (pbk)
ISBN: 978-1-315-71463-9 (ebk)

Acquisition Editor: Wendy Fuller
Editorial Assistant: Grace Harrison
Production Editor: Ed Gibbons

Typeset in Univers
by Saxon Graphics Ltd, Derby

Printed and bound in Great Britain by
TJ International Ltd, Padstow, Cornwall

CONTENTS

FOREWORD

ONE OF THE central challenges of the 21st century is designing intelligent forms of human settlement. In the last 200 years, the global population grew from 1 billion to 6.9 billion. In less than four decades, the global population will reach 9 billion. Our highly consumptive pattern of development that relies upon an inexhaustible supply of arable land, water, and energy cannot be sustained. Buildings and transportation today create two-thirds of the carbon in our atmosphere. Where we place our buildings, the way we build them, and the way in which we move between them are the major causes of climate change. The future of our planet is contingent upon our ability at the beginning of this new millennium to create cities of delight—urban environments that are dense, compact, and highly livable.

A key to achieving this is an American invention: the tall building. The experiment in vertical building begun by Jennings, Burnham, Sullivan, and others in Chicago after the Great Fire of 1871 is central to the long-term sustainability of our planet. The earliest examples of this building form, begun in earnest in the 20th century, must now be reconsidered with the understanding that the tall building is not simply an expression of corporate power or civic pride—it must become the very basis of human settlement.

To physically realize compact, dense, and humanistic vertical cities in a future of limited material resources, radical innovations in architecture and structural engineering are necessary. Mark Sarkisian and his colleagues at SOM have embraced this challenge. Their process begins not with columns and beams, but rather with an intuitive understanding of the interrelationship between forces at play. They go beyond the rote, normative process of structural optimization to consider new ways of achieving holistic building efficiencies in which the essence of architectural form becomes part of the solution.

Mark advocates an intuitive and organic understanding of structure and form in SOM's design studios. To support this nuanced approach, he and his colleagues have assembled extraordinary new tools to achieve innovation. These systems combine computational analysis with visualization models similar to those used by scientists to quickly visualize organic form and behavior at the molecular level. This work is undertaken in a collaborative, multidisciplinary environment, much like the "group intelligence" model of today's collaborative sciences.

Mark and his colleagues have extended this collaborative model beyond SOM's professional studios to the academic studio. SOM's engagement with leading universities in many disciplines over the years has yielded constant reciprocal benefits—inspiration and research—for the firm and the academy. This book, *Designing Tall Buildings*, is a direct result of the engineering and architecture class at Stanford University that Mark began with architect Brian Lee in October 2007. It is a book that will no doubt provide inspiration as well as practical guidance to students, professors, practicing architecture, and engineering professionals, and design devotees.

Craig W. Hartman, FAIA
January, 2011

INTRODUCTION

THIS BOOK IS meant to illuminate the design process for tall building structures with fundamental concepts and initial considerations of the site developed into complex solutions through advanced principles related to natural growth and the environment. The author's goal is to give a holistic description of all major considerations in the structural engineering design process. Specific examples of work developed at Skidmore, Owings & Merrill LLP (SOM) are used for each step in the process.

The work within this book represents decades of development by the architects and engineers of SOM. As an integrated practice, the work is the product of a close design collaboration—leading to many innovations and particularly in tall building design. Pioneering structural systems including adaptation to form, material efficiencies, and high performance have resulted from this work.

The catalyst for this book came from the need for structural engineering curriculum in a SOM-led Integrated Design Studio class at Stanford University. The goal was to teach architectural and structural engineering design, in parallel and with equal emphasis, focused on tall building design challenges which included complex programmatic and site considerations. Each chapter in this book was developed as a class lecture, focused on a particular subject in the design process.

The book begins with a select history of tall buildings, the inspiration behind their designs as well as some early analysis techniques for their design. Fundamental principles for tall building design are developed with the premise that these structures should be designed and built considering simplicity, structural clarity, and sustainability. The site is considered for geotechnical, wind, and seismic conditions. Forces from gravity and lateral loads including wind and earthquakes are described with load combinations required for design. Multiple codes are referenced to offer different approaches to the calculations of loads on the structure.

The language of the tall building structure is described to provide a better understanding of major structural components and overall systems. Framing diagrams from buildings are included to describe the use of major structural building materials such as steel, concrete, and composite (a combination of steel and concrete). Attributes of tall structures are described including strength and serviceability—building drift, accelerations, and damping—among others. Tall building characteristics such as dynamic properties, aerodynamics associated with form, placement of materials, and aspect ratios are also described.

Suggested structural systems based on height and materials are reviewed. These systems result in the greatest efficiency through least material when considering gravity and lateral loads. Inspirations from nature through growth patterns and natural forms are considered for the development of more advanced ideas for tall building structural systems. Natural behavior is contemplated through the considerations of structure behaving mechanically rather than statically when subjected to load—particularly in seismic events. Correlations to mathematical theories such as the Fibonacci Sequence and genetic algorithms, as well as the use of Emergence Theory in structural design, are considered.

Performance-based design has become an important approach to the design of tall buildings where exceptions to codes are made. A special procedure for considering this non-prescriptive design method is given. Finally, and perhaps most importantly, considerations for effects on the environment including embodied energy and equivalent carbon emissions are contemplated.

DESIGNING TALL BUILDINGS

STRUCTURE AS ARCHITECTURE

CHAPTER 1

PERSPECTIVE

1.1 HISTORICAL OVERVIEW

THE FIRE OF 1871 devastated the City of Chicago but created an opportunity to rethink design and construction in an urban environment, to consider the limits of available, engineered building materials, to expand on the understanding of others, and to conceive and develop vertical transportation systems that would move people and materials within taller structures.

In the late 1800s technological advancements led to the development of cast iron during the United States' industrial revolution. Although brittle, this material had high strength and could be prefabricated, enabling rapid on-site construction. The first occupied multi-story building to use this technology was the Home Insurance Building located in Chicago. Built in 1885 with two floors added in 1890, it was 12 stories tall with a height of 55 m (180 ft). Though it has since been demolished, it is considered the first skyscraper.

Chicago in Flames—The Rush for Lives Over Randolph Street Bridge (1871), Chicago, IL

The Chicago Building of the Home Insurance Co., Chicago, IL

FACING PAGE
Willis Tower (formerly Sears Tower), Chicago, IL

The Monadnock Building, Chicago, IL

Monadnock Building Detail at Base, Chicago, IL

The 16-story Monadnock Building located in Chicago and constructed in 1891 used 1.8 m (6 ft) thick unreinforced masonry walls to reach a height of 60 m (197 ft). The structure exists today as the tallest load-bearing unreinforced masonry building. The 15-story, 61.6 m (202 ft) tall Reliance Building built in 1895 used structural steel and introduced the first curtain wall system. Buildings now could be conceived as clad structural skeletons with building skins erected after the frame was constructed. The Reliance Building has changed use (office building converted to hotel), but still exists on State Street in Chicago. Steam and hydraulic elevators were tested for use in 1850. By 1873, Elisha Graves Otis had developed and installed steam elevators into 2000 buildings across America. In 1889, the era of the skyscraper was embraced with the first installation of a direct-connected, geared electric elevator.

Identity and egos fueled a tall building boom in the late 1920s and early 1930s with other urban centers outside of Chicago getting involved. In 1930, the Chrysler Building in New York became the world's tallest, with the Empire State Building soon to follow. Completed in April 1931 (completed in one year and 45 days), at 382 m (1252 ft), it surpassed the Chrysler Building by 62.2 m (204 ft). The total rental area in the tower is 195,000 square meters (2.1 million square ft). The most significant feat was the extraordinary speed in which the building was planned and constructed through a close collaboration between architect, engineer, owner, and contractor.

Reliance Building, Chicago, IL, Left—Steel Frame, Right—Completed Building

The Empire State Building's first signed contract for architectural services with Shreve, Lamb, and Harmon was in September 1929, the first structural steel column was placed on 7 April 1930, and the steel frame was topped off on the eighty-sixth floor six months later (the frame rose by more than a story a day). The fully enclosed building, including the mooring mast that raised its height to the equivalent of 102 stories, was finished by March 1931 (11 months after the first steel column was placed). The opening day ceremony took place on May 1, 1931. The structural engineer, H.G. Balcom (from a background in steel fabrication and railroad construction), worked closely with general contractors Starrett Brothers and Eken to devise a systemized construction process.

Three thousand five hundred workers were on-site during peak activity. Some 52,145 metric tonnes (57,480 tons) of steel, 52,145 cubic meters (62,000 cubic yards) of concrete, 10 million bricks, 6,400 windows,

Beaux Arts Architect Ball (1931), New York, NY

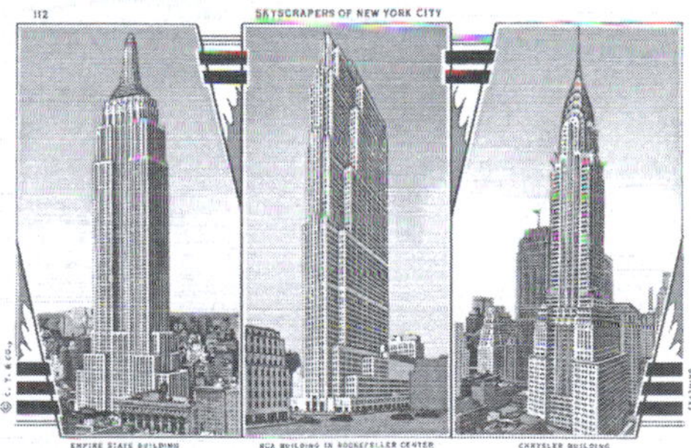

Empire State Building, RCA Building in Rockefeller Center, Chrysler Building, Skyscrapers of New York linen postcard (1943)

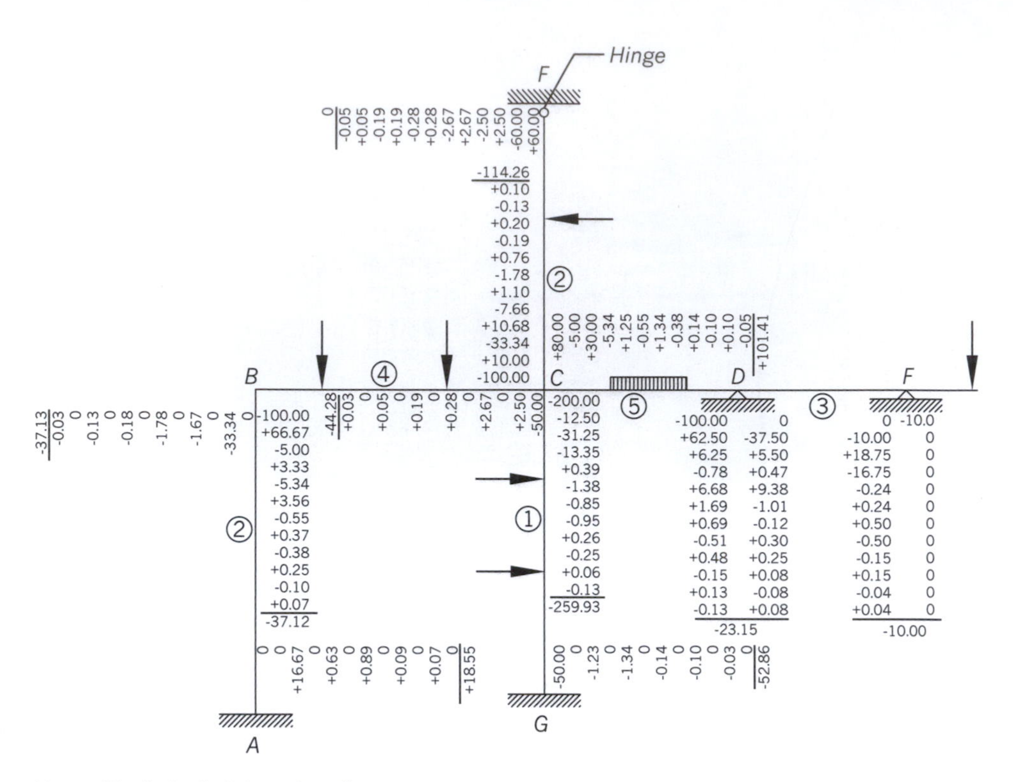

Moment Distribution for Indeterminate Structures

and 67 elevators were installed. The Empire State Building remained the tallest building in the world for 41 years until the World Trade Center in New York was built in 1972. The Chrysler, Empire State, and World Trade Center Buildings were all constructed of structural steel.

The development of more sophisticated hand calculation techniques for structures, including methods developed by great engineers like Hardy Cross, allowed engineers to analyze, design, and draw structures that could be easily constructed. Urged by the University of Illinois' Dean of Engineering Milo Ketchum, Cross published a ten-page paper titled "Analysis of Continuous Frames by Distributing Fixed-End Moments" in 1930, showing how to solve force distribution in indeterminate structures, which was one of the most difficult problems in structural analysis.

World War II temporarily halted homeland construction because of the need for steel products in the war efforts. It wasn't until the late 1950s and early 1960s that interest in tall buildings was renewed. Great architects such as Mies van der Rohe used structural steel to create a minimalistic architectural approach. His notable tall building projects included 860 & 880 North Lake Shore Drive (1951) and 900 & 910 North Lake Shore Drive (1956) in Chicago. Skidmore, Owings, & Merrill (SOM) developed building designs

CLOCKWISE FROM TOP LEFT

Lever House, New York, NY

Inland Steel Building, Chicago, IL

One Bush Street (formerly Crown Zellerbach), San Francisco, CA, 1959
(photograph by Morley Baer. ©2015 by the Morley Baer Photography Trust, Santa Fe. Used by permission—All reproduction rights reserved)

One Maritime Plaza (formerly Alcoa Building), San Francisco, CA, 1967
(photograph by Morley Baer. ©2015 by the Morley Baer Photography Trust, Santa Fe. Used by permission—All reproduction rights reserved)

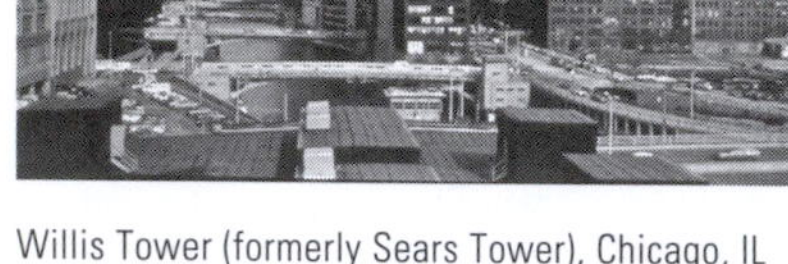

Willis Tower (formerly Sears Tower), Chicago, IL

John Hancock Center, Chicago, IL

that used structural steel to create long, column-free spans allowing for flexible open office spaces while creating a corporate identity through the finished building. These projects included The Lever House—New York (1952), the Inland Steel Building—Chicago (1958), the Crown Zellerbach Building/One Bush Street—San Francisco (1959), and the Alcoa Building—San Francisco (1964).

It wasn't until the late 1960s and early 1970s that considerable new development in tall building analysis, design, and construction were made. The Cray Computer provided the analytical horsepower to evaluate buildings such as the John Hancock Center (1969) and the Sears Tower (1973) located in Chicago. Prefabricated, multi-story, modular building frame construction was used to reduce construction time. Wind engineering, largely developed by Alan Davenport and Nicholas Isyumov at the University of Western Ontario, provided vital information about the performance of buildings in significant wind climates. Geotechnical engineering, led by engineers such as Clyde Baker, provided feasible foundation solutions in moderate to poor soil conditions. Probably the most important contribution was SOM's late partner Dr. Fazlur Khan's development of economical structural systems for tall buildings. His concepts were founded in fundamental engineering principals with well-defined and understandable load paths. His designs were closely integrated with the architecture and, in many cases, became the architecture.

Brunswick Building, Chicago, IL

Chestnut-Dewitt Tower, Chicago, IL

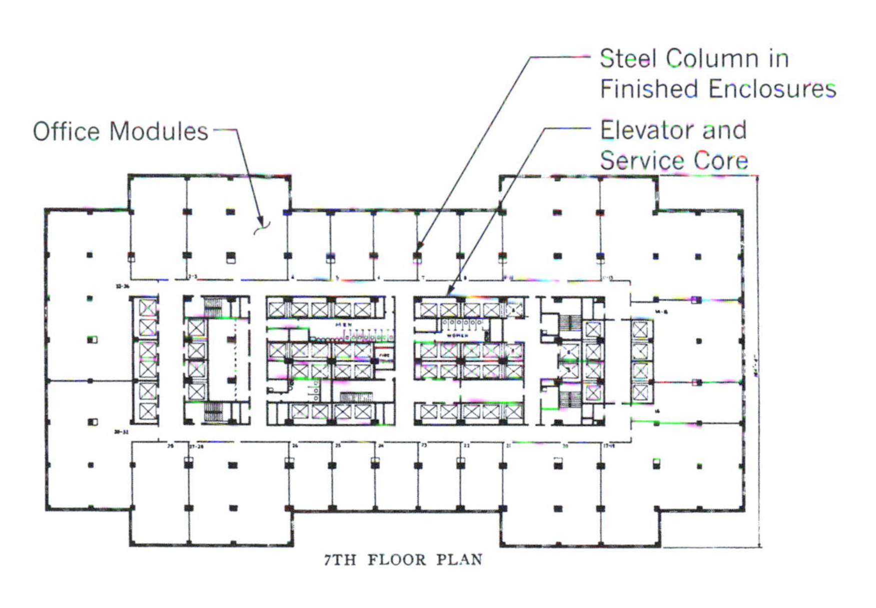

Empire State Building, Typical Floor Plan, New York, NY

Drawing from *Fortune Magazine*,
September 1930, Skyscraper Comparison

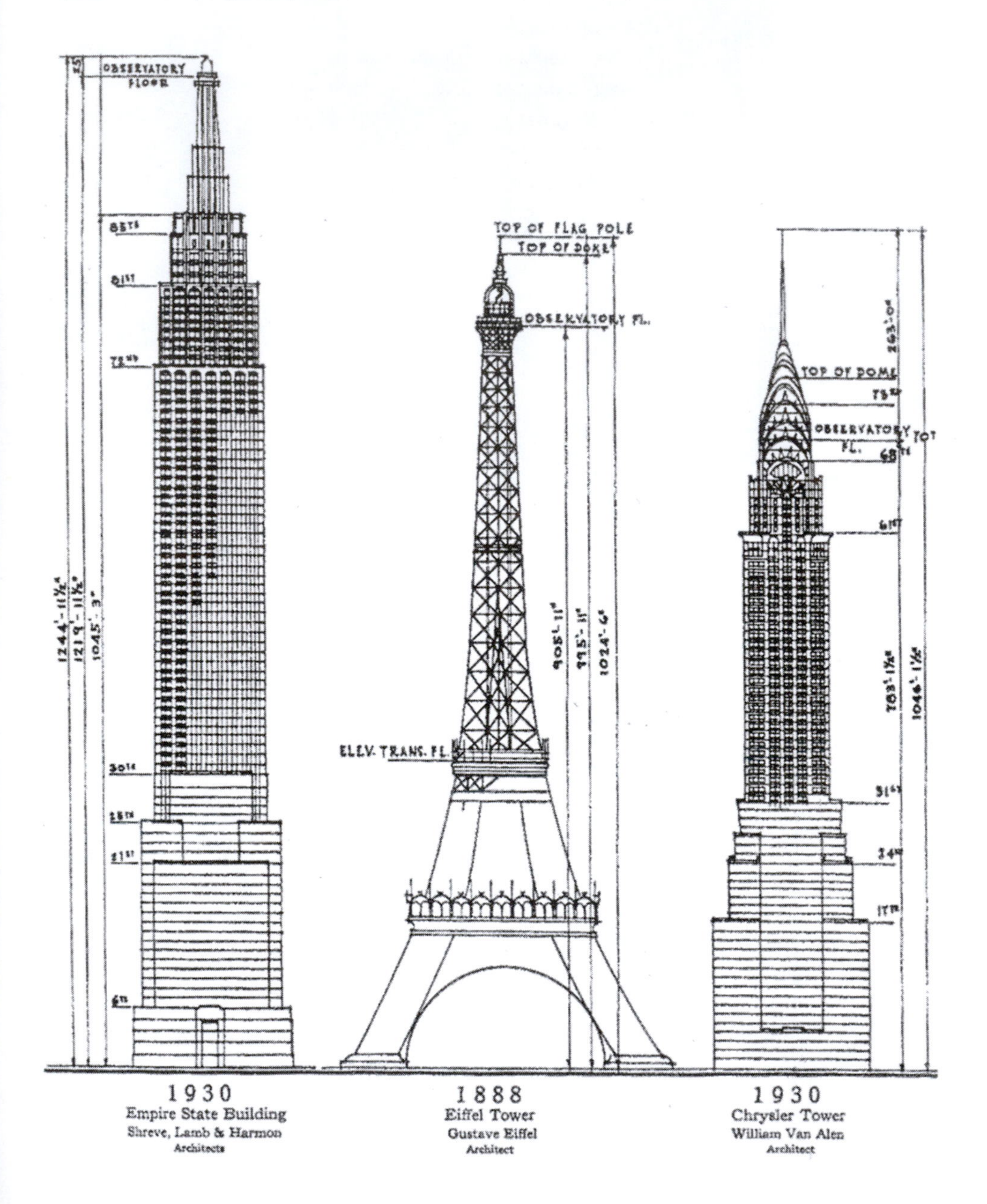

At the same time, SOM also developed tall building structural system solutions in reinforced concrete. An increased understanding of concrete's chemical and physical characteristics combined with consistently higher compressive strengths led to an economical alternative to structural steel in tall building structures. The Brunswick Building (1964) and Chestnut Dewitt Tower (1965) located in Chicago were major structures that used this technology.

Tower height depends on material strengths, site conditions, structural systems, analytical/design capabilities, the understanding of building behavior, use, financial limitations, vision, and ego. Concepts in tall building evolve rather than radically change. Increased understanding of material combined with greater analytical capabilities have led to advancements. Computer punch cards and Cray Computers have been replaced by laptops with comparable computing capabilities. Many advancements are based on architectural/engineering collaboration.

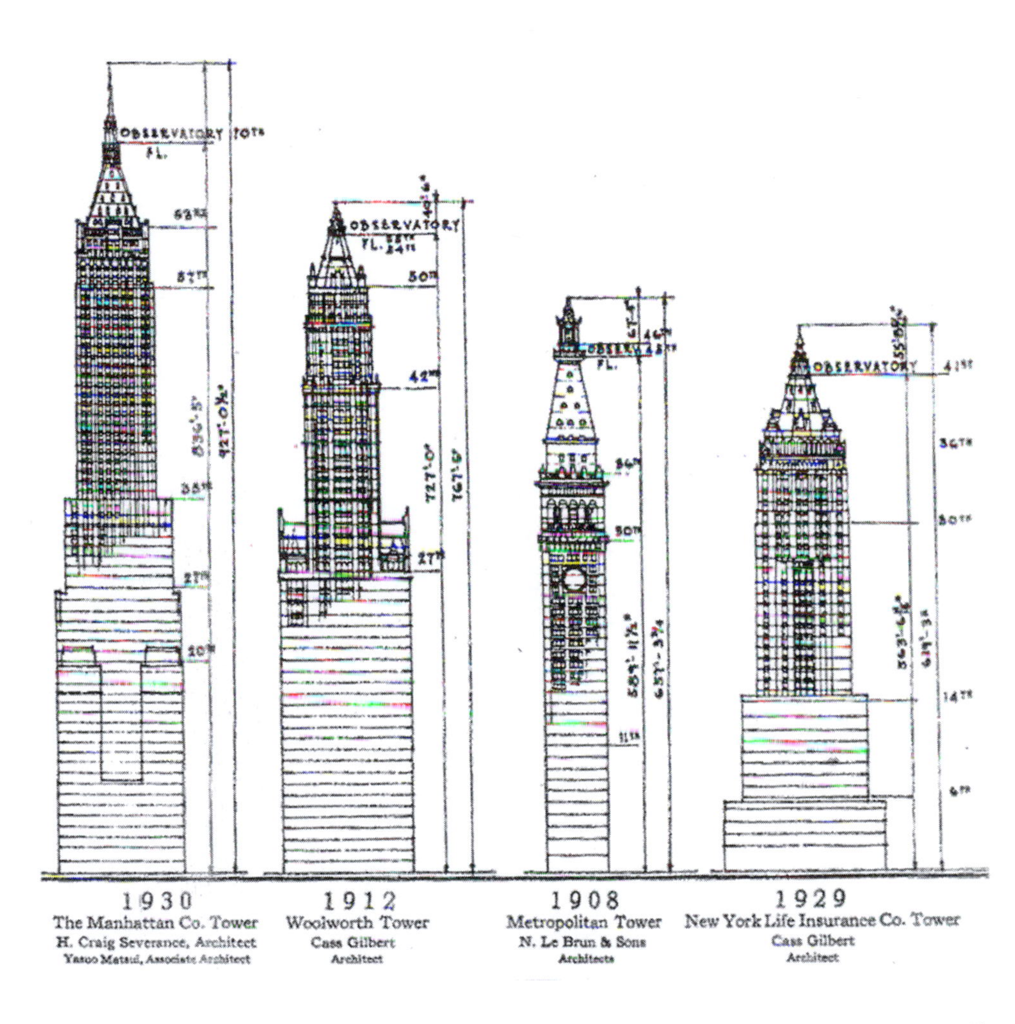

1.2 FIVE AGES OF THE SKYSCRAPER

Renewal and opportunity followed the Great Chicago Fire, first through Chicago's World's Columbian Exposition scheduled for completion in 1892 (four centuries after the discovery of the Americas by Columbus) but finished in 1893 with the development of new ideas from Daniel Burnham, William Holabird, Louis Sullivan, John Wellborn Root, and others. Following the exposition, the first skyscraper age as well as the era of the First Chicago School emerged corresponding with the early use of structural steel and advances in structural engineering. The age ushered in the idea of structural frames clad with exterior wall systems and vertical transportation through passenger and freight elevators. The first skeletal form with a glass and structured façade was designed by William Le Baron Jenny, a civil engineer who practiced as an architect and was considered the father of the First Chicago School, through the Leiter Building (1879), and later in the Home Insurance Building (1885), which is considered the world's first skyscraper. Jenny, Sullivan, and Root, among others, designed structures that were utilitarian, economical, and free of excessive ornamentation. Root's Monadnock Building completed in 1891 is an excellent example of a structural response to force flow through gradually increasing widths and depths of the masonry walls as the building meets its foundations.

Rendering of Super-Frame from Myron Goldsmith's Master's Degree Thesis

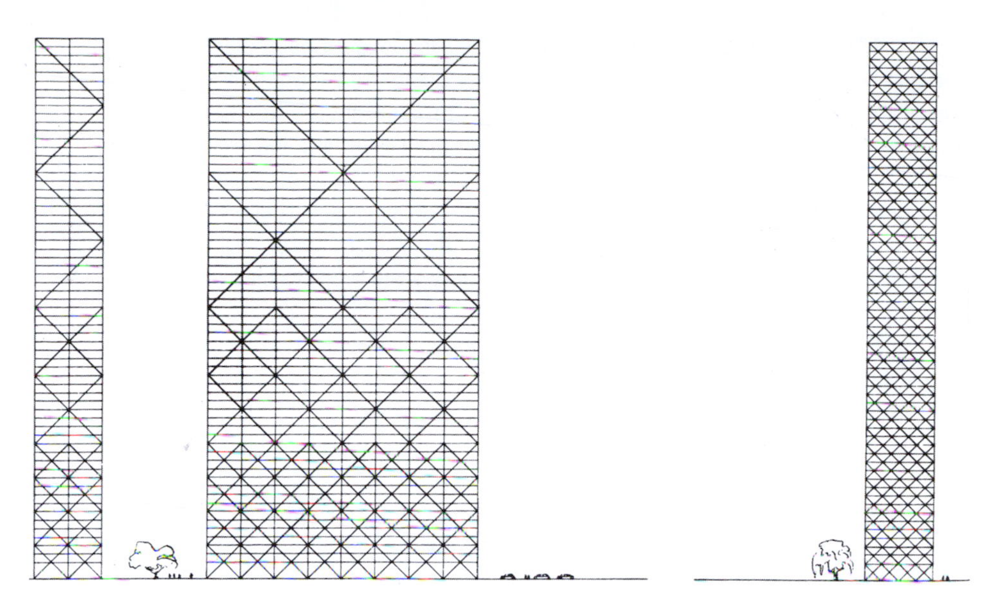

Rendering of Braced Frames from Myron Goldsmith's Master's Degree Thesis

The second skyscraper age was one that used steel construction to new heights while seeking aesthetic inspiration from classic historic models including style and ornamentation from Greek and Roman monuments. Especially in New York City, corporate-owned skyscrapers became a symbol of strength and prosperity. The Chrysler, Empire State, and Rockefeller Center Buildings are important symbols of this period.

After a sharp decline in construction in the period leading up to and following World War II, construction of tall buildings began again in the late 1950s. The third skyscraper age and the formation of the Second Chicago School were dominated by European architects such as Mies van der Rohe and Le Corbusier. Heavier masonry façades were replaced with metal and glass, and art-deco aesthetics were replaced with the International Style which emphasized the expression of structure by exposing it on the exterior of the building. Myron Goldsmith, a student of Mies, became an important contributor to this modernist movement. In his master's thesis "The tall buildings and the effects of scale," Goldsmith considered structure with varying modules and configurations contemplating rational force flow and structural efficiency. Recognizing that a super-frame used for a tall building lateral system lacked stiffness to resist all forces on its own, Goldsmith understood that a secondary infill frame could be used to enhance the stiffness of the system. He also understood that a braced frame could be beautifully transitioned over the height of a building structure increasing in density where lateral load demands were highest.

Many buildings designed in the late 1970s lacked a particular style and recalled ornamentation from earlier buildings designed during the second age of the skyscraper. The fourth age of the skyscraper ignored the environment and loaded structures with decorative elements and extravagant finishes. Sculptural imagery and monumental expression dominated architectural practices bored with modernism. Fazlur Khan was one that strongly opposed this approach to design and considered the designs to be whimsical rather than rational. Most importantly he considered the work to be a waste of precious natural resources.

Khan's scathing assessment of postmodernism was summarized in his written address to the Architecture Club of Chicago in 1982, accepting his unprecedented election to president of the club. Khan wrote

> Today it seems the pendulum has swung back again towards architecture that is unrelated to technology and does not consciously represent the logic of structure. Nostalgia for the thirties and even earlier times has hit a large segment of the architectural profession; in many cases façade making has become the predominant occupation. It is apparent that postmodernism in architecture is very much the result of the architect's lack of interest in the reality of materials and structural possibilities: the logic of structure has become irrelevant once again. This attitude in architecture suits many engineers because of their overspecialization in engineering schools which treat the solution of the problem as the ultimate goal, and not the critical development of the problem itself.

Khan's address to the Architectural Club in 1982 ended with signs of optimism. He concluded "but logic and reasoning are strong elements of human existence, always important when man must transcend into the next level of refinement. There are already some signs of that happening in architecture. New structural systems and forms are beginning to appear once again and with them new architectural forms and aesthetics. The pendulum of structural logic in architecture continues to swing." Khan died before the address could be given.

Since Khan was deeply interested in technology and its effects on society, the developments of his work led us to a more comprehensive consideration for structures that will define the next age of the skyscraper. The next, and fifth age of the skyscraper, will focus on the environment including performance of structures, types of materials, construction practices, absolute minimal use of materials/natural resources, embodied energy within the structures, and perhaps most importantly, a holistically integrated building systems approach which will shape this movement.

These buildings will be designed to respond to any loading without damage, and regeneration rather than only consumption will begin to appear.

Baietan Urban Design Master Plan in Guangzhou, China

LEED (Leadership in Energy and Environmental Design) is only a start in raising awareness of responsible design approaches and other ideas will develop with requirements specific to limits of embodied energy for buildings or financial incentives for those that reduce them.

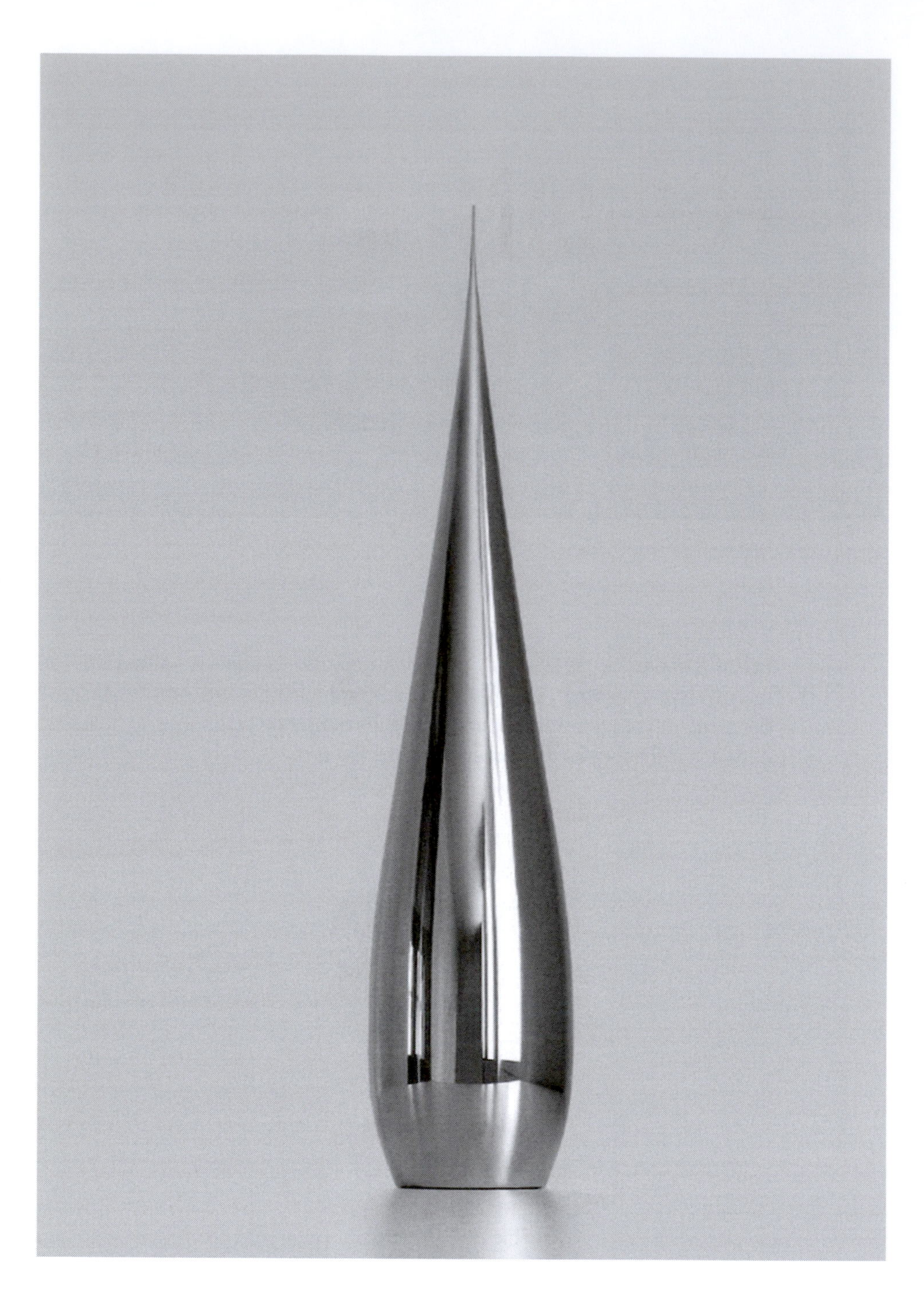

Tower Shape Optimized for Wind Effects

CHAPTER 2
FUNDAMENTALS

THE FUNDAMENTALS OF tall building design, particularly those related to structures that define architecture, are based on an integrated approach that considers science, application of science (engineering), productive use of space, conservation of natural resources, and long-term value. Sustainable architecture is founded in engineering concepts that are innovative whether the structure is implicit in the architecture or explicitly expressed.

2.1 GUIDING PRINCIPLES

The world is facing dire challenges of climate change and the depletion of natural resources as a result of unsustainable population growth and industrialization. Simplicity, structural clarity, and sustainability not only define a visual quality for buildings, but also form the guiding principles for tall building design. It is the development of these three principles that leads to responsible design and construction of towers the can impact climate change and reduce demands on resources.

2.1.1 Simplicity

Simplicity in form and purity of concept are perhaps the most important fundamentals for efficient tall building design. Although not always possible, symmetry, uniformity of mass, and control of force flow within the structure leads to least use of material. Using fundamental theories of physics to design for system behavior results in the management of/coexistence with natural and/or human-made environmental conditions and informs placement of material where best suited. The form of the structure must be developed considering several variables including program, height restrictions, soil conditions, potential imposed lateral loads, as well as relationships to other structures to determine daylight exposure, among other reasons. Purity of concept includes a structural engineering response that is sympathetic to architectural goals: structure in direct response to architecture, absent of frivolous materials and with the systems having multiple purposes. Simplicity of form and purity of concept are typically interpreted and developed into new ideas that further enhance performance. Examples of this include the moment frame developed into a tubular frame, and a tubular frame developed into a braced tubular frame to address increases in height without substantial increases in structural materials.

2.1.2 Structural Clarity

Structural clarity may be visually understood, but most importantly must define a system with clear load paths. Creating certainty in environments with potentially very uncertain events is the goal of this attribute. For instance, seismic ground motions including direction, magnitude of accelerations, and displacements during earthquakes are highly uncertain. Layered soil conditions beneath tall building structures further complicate the loading state during the seismic event. A load resisting system where force flow is clearly understood creates certainty in environments prone to uncertain events. Ground motions related to seismicity are among the loading events with the greatest uncertainty of magnitude and direction.

Business Men's Assurance Building, Kansas City, Missouri

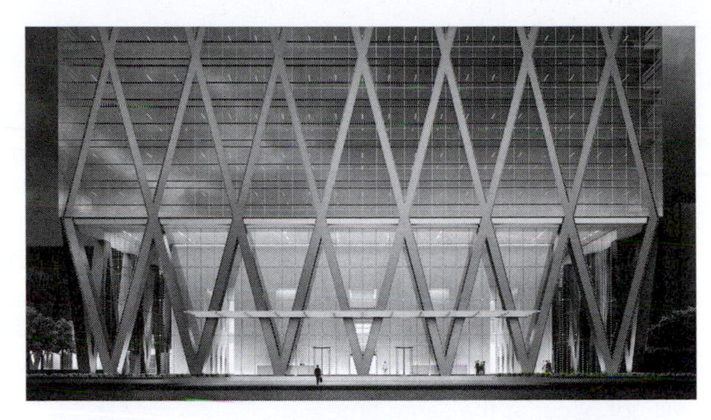

Base of Tower, Guohua Financial Tower, Ningbo, China

Stack Effect in Tall Buildings

2.1.3 Sustainability

Ultimately, tall buildings should be designed and constructed to be self-sufficient, if not regenerative. The structures, or local districts where these structures exist should be designed to generate power to fully operate, capture rainwater, treat and reuse wastewater, and even produce food.

Tall building structures must be resilient, incorporating systems that enhance performance and extend life. These structures should be designed considering a life expectancy that is extended by high performance architectural, building services, and structural systems. To the greatest extent possible, structures should include recycled materials and be supplied locally. Since concrete is the most heavily used building material worldwide, goals should include the elimination of carbon-emitting cement by replacing it with admixtures that can be combined with slag. Every major building component should be designed for at least two purposes. For example, structures should be designed to control heat gain, exterior wall systems should be designed to not only control the indoor environment but also generate electricity, and foundation systems should be used to circulate water through geothermal techniques, allowing major mechanical systems to benefit from reduced demand.

2.2 DESIGN LOADS

Design loads for tall building structures can be considered as sustained and transient. Gravity loads are considered sustained, whereas wind, seismic, temperature, and snow are considered transient. Self-weight of the structure is permanent and has the highest level of certainty where superimposed loads very depending on the system. Exterior wall loads are perhaps the most certain, but partitions, ceilings, and mechanical systems may be less certain. Load combinations that combine gravity and transient loads include reduction factors to account for the probability that all of the maximum loads will not occur at the same time.

2.2.1 Gravity

Proportioning gravity loads into vertical supporting elements is important to control the behavior of a tall building structure. These loads can be used as ballast, reducing or eliminating overturning effects. This ballast is an important consideration for foundation design especially where systems have limited uplift capacity. Furthermore, rotations at the base of a tall tower translate into large movements at the top of a tall building. Even distribution of sustained gravity loads is important to reduce or eliminate differential settlement that could also result in large displacements and the top of the structure. More

Column Sizes Responding to Increased Gravity Load, One Shell Plaza, Houston, TX

important could be the impact of gravity loads located eccentrically to the foundation where overall bending moments could occur due to a P-delta phenomenon. Vertical structural elements such as columns and walls need to be designed for the increased load over the height of the building with dimensions or material strength, or both, increasing as load increases toward the bottom of the structure.

2.2.2 Wind

The effects of wind exist for all tall building sites. The magnitude, predominant wind directions, and impact of neighboring terrain vary for all sites. Building codes address wind conditions based on historic climatic data and base formulaic approaches to defining loads based on theory and research. Since the wind environment can be complex, wind tunnel modeling is important for taller structures with the boundary layer modeled and impacts of existing and perhaps new structures planned for the future.

Strength and serviceability are the two primary considerations. Typically tall structures are designed for strength in accordance with appropriate codes and standards for a 100-year return period. Longer return period winds are commonly considered for stability. Tall building structures are evaluated for 1000-year or even greater events to confirm stability with actual expected structural material strengths and no additional load amplification factors included. A 50-year return period is typically considered for serviceability, primarily related to building drift; however, more frequent return periods are considered for evaluating perception to motion. These more frequent events are typically considered for 1 and 10-year return periods.

Perhaps the most important consideration for wind effects is the structure itself. Shaping the tower, providing openings in the façade, and

Opening in Façade to Reduce the Impact of Wind Load, Poly Real Estate Headquarters, Guangzhou, China

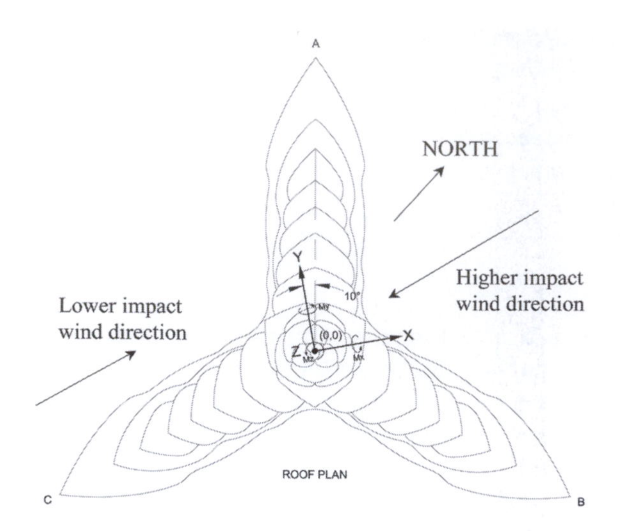

Stepping of Façade and Tower Orientation Used to Minimize Wind Effects, Burj Khalifa, Dubai, UAE

orientation on the site all greatly impact the wind load demands on the tower and can lead to significant reductions in materials required to resist wind for both strength and serviceability.

The Burj Khalifa was carefully optimized for wind. The stepping of the building façade and the placement of the tower on the site was optimized to reduce wind load demand. Wind tunnel modeling revealed that winds subjected to the noses of the tower (winds applied to points A, B, and C in the figure) of the tower had less of an impact on the structure because of cut water effects than winds subjected to the tail of the tower (winds applied to areas in between points A, B, and C in the figure). The tower was placed so that the majority of predominant winds would be applied to the tower at the noses.

The stepping of the tower was designed to disrupt vortex shedding along the height of the tower. These vortices are not organized and vary over the height of the building, lowering the impact of wind. It was determined that due to these measures, and the inherent mass and stiffness of the structure, that no supplemental damping was required.

Contrary to common thinking, tall building structure design is typically governed by across-wind motion rather than along-wind motion. This phenomenon is due to the dynamic effects of wind on towers and the normal forces that are applied to the perpendicular sides of towers when winds are applied and vortex shedding is developed. Wind pressures increase with height from the bottom to the top with forces in the structure accumulating and increasing from the top of the tower to the bottom.

Resonance could be experienced in a tall tower due to modes of vibration with high frequencies that correspond to wind gusts combined with low levels of damping.

Wind Tunnel Testing, Burj Khalifa, Dubai, UAE

2.2.3 Seismic

Forces within the tall building structure developed from seismic ground motions are based on the fundamental equation of physics, F = mass x acceleration. Inertial forces are generated within the structure and accumulate with the highest demands on the structure at its base. Seismic mass of the tower is fairly straightforward to calculate, but the acceleration used for design is more complex.

Determining the expected ground acceleration for a particular site is only the first step in determining acceleration. The dynamic characteristics are important. If the structure was infinitely stiff, the structure would experience a maximum force defined by the product of mass times the acceleration of the ground. The maximum acceleration of the ground in a given direction is called the peak ground acceleration (PGA) meaning the mass would move with the same motion of the ground. If the structure was infinitely flexible with no stiffness, then the mass of the structure would remain stationary relative to the motion of the ground and no forces would be generated in the tower. All structures fall somewhere in between these two extremes. Typically structures with greater flexibility attract less force from the ground than those with high stiffness. Therefore, tall building structures, when designed properly for the anticipated movement, behave much better in an earthquake than a short stiff structure.

The other factors that influence seismic demand on structures are soil characteristics, ductility, and damping. Soft soils magnify ground motions and increase demand on towers. Ductility is related to the amount of energy dissipation in the structure during a seismic event. The higher the ductility in the system, the less force the system experiences. Therefore, a tower that incorporates a flexible moment-resisting frame where behavior is governed by bending of members will see less force than a diagonally braced frame where members are typically subjected to axial loads. Finally, damping is key to the amount of acceleration and, consequently, forces that the structure will experience. The higher the damping within the structure, the lower the forces the structure will experience.

Typically the fundamental period of vibration will control the behavior of the structure particularly for structures with a period of 2 seconds or less corresponding to approximately 20 stories or fewer. Typically the fundamental period of vibration corresponds to lateral translation of the structure when excited and regular in shape and mass, but could correspond to torsion if the structure is irregular. The fundamental period is the time it takes a structure to move through one full cycle of motion. For tall towers, higher modes of vibration code create significant forces due to the activation of mass associated to high frequencies.

2.3 FORM AND RESPONSE

Tall building structures are especially susceptible to environmental conditions. Typically a tapered tower will behave more favorably than a tower with constant plan dimensions over the height when subjected to significant winds. The area of the tower high above the foundation that would need to resist wind forces is reduced; therefore, the bending moment at the base of

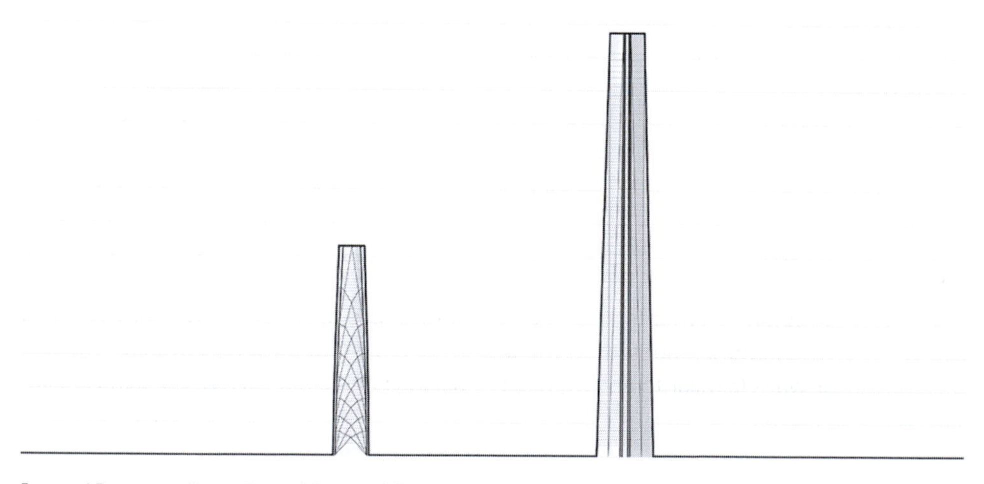

Form and Response—Proportion and Structural Systems

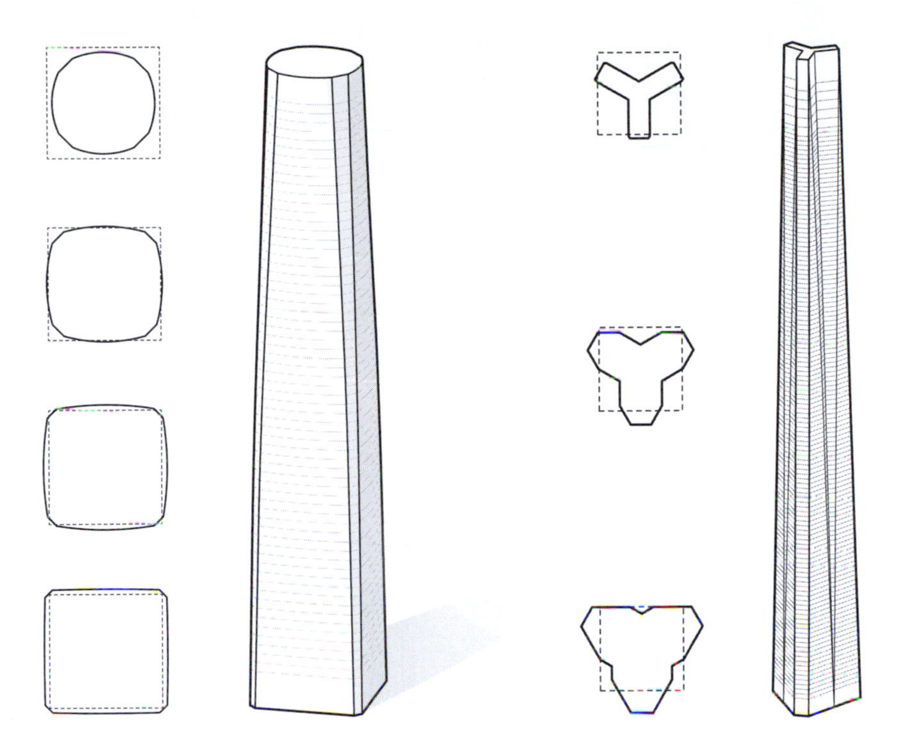

Tapered Form and Transitioning Square Plan Shape

Tapered Form and Transitioning Triangular Plan Shape

the tower is significantly reduced since bending moment = force times distance. This tapering also typically reduces seismic demand since less mass exists far away from the base support. The structural system for a tower with a high aspect ratio (ratio of height to least structural dimension) may be proportioned differently.

Structural systems for these tower forms will vary and respond to program area requirements and height limits. For instance, the shorter tower is proportioned to maximize floor area per floor where the taller tower has less area per floor and a much taller height limit. The structural system for the shorter tower can take advantage of proportionally longer plan dimensions relative to its height where an optimized frame system can be used to manage lateral loads. In the taller, more slender tower, vertical structural elements are much more effective in resisting overturning loads when interconnected to the central core system through outrigger trusses or walls (similar to an arm used to balance one's stability while using a cane to walk).

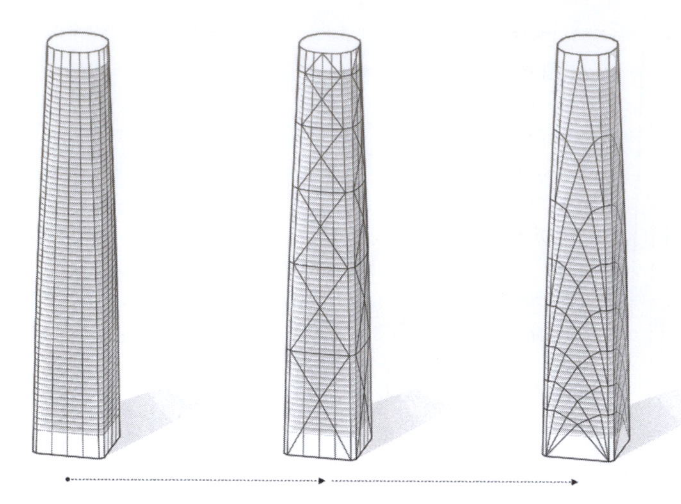

Evolution of Structural System: Tubular—Braced—Optimized Frame

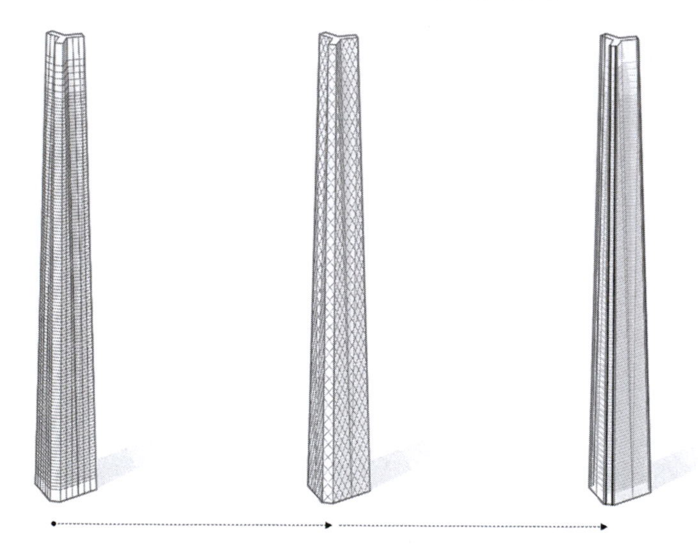

Evolution of Structural System: Tubular Frame—Mesh Frame—Mega-Column

2.4 STIFFNESS AND SOFTNESS

The dilemma in tall building design is developing systems that respond to multiple loading types and balance stiffness and softness. Frequently tall buildings are designed and built in regions susceptible to high wind loads, as well as high seismic loads. The structural system must be stiff enough to control displacements and accelerations when subjected to wind loads. For a tall building, the most efficient design eliminates all bending of individual members, and the best way to achieve this is through a diagonal bracing system. This same system must behave favorably in a seismic event where loads could be many times greater than wind loads. Stiffness attracts force in structure. Therefore, the best system is one with softness or ductility. Ductility can be achieved with a state change in materials, for instance in steel that is forced to permanently deform under load (become plastic). A far superior response is to create plasticity through fusing of systems, perhaps by the dissipation of energy through mechanical devices such as friction fuses.

The fundamental period (T) of the building, the natural frequency associated with free vibration, is the key to understanding stiffness and softness.

$$T = 2\pi \sqrt{M/K}$$

where:

M = mass of the structure
K = stiffness of the structure

For a structure with the same mass, a reduction of stiffness increases the period. Structural systems in tall buildings that can change stiffness are the key to good behavior, adaptability, and resilience when subjected to both types of lateral loads.

2.5 MATERIALS

Materials used within tall buildings should be evaluated based on multiple factors, in addition to life safety. Since the use of materials accounts for the single largest contributor to emitted carbon in building construction, considerations for appropriate and efficient use of these materials is paramount. In addition to material-based carbon, considerations for construction speed, flexibility for future modifications, self-weight, framing depth, and even mass specific gravity are important (storage of energy within the structure).

Horizontal framing systems within tall building structures are important in determining total material use. For instance, a framing system with less mass will result in less seismic demand on a structure compared to one that is more massive, assuming the lateral stiffness of the structure is the same. Floor framing with less depth results in less materials for exterior wall and mechanical systems given the same ceiling height on occupied floors.

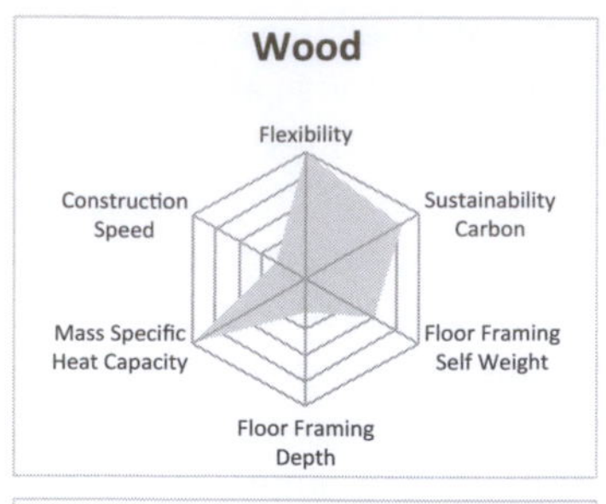

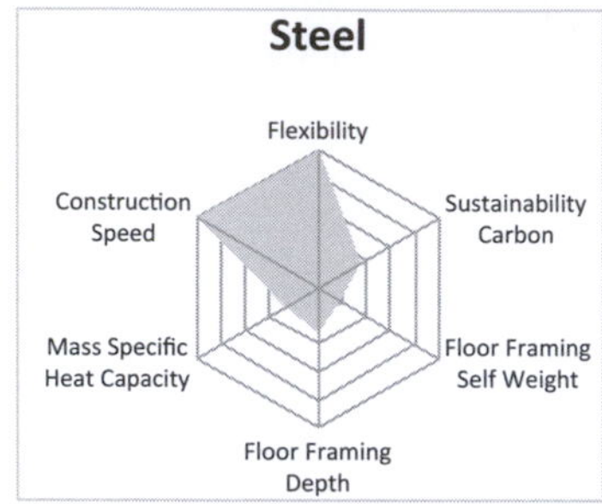

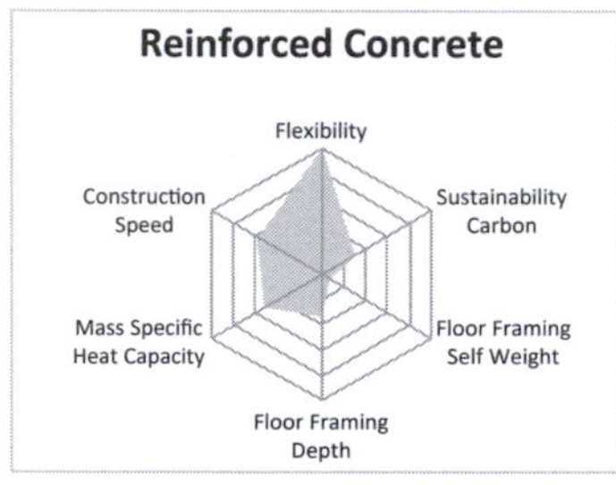

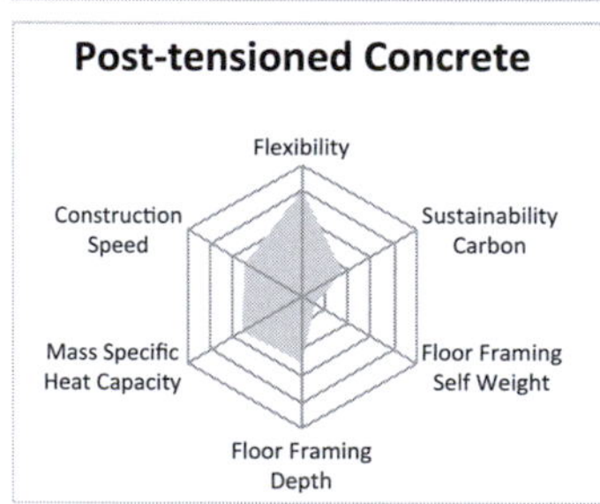

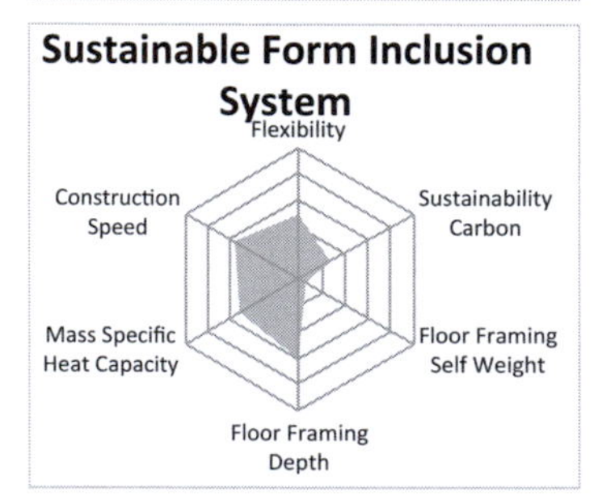

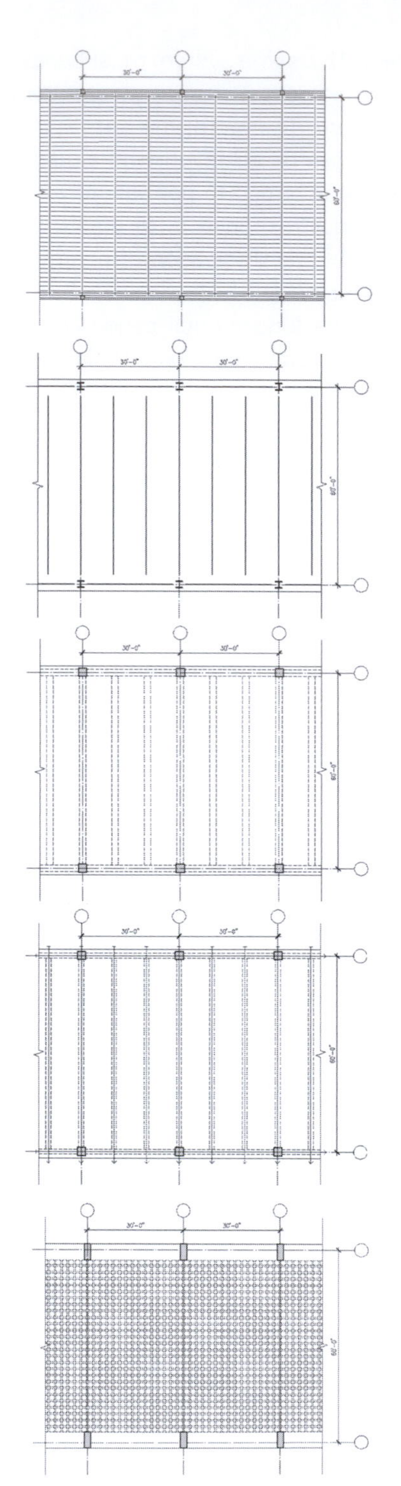

Multi-Variable Evaluation of Framing Systems [18 m × 9 m (60 ft × 30 ft) module]

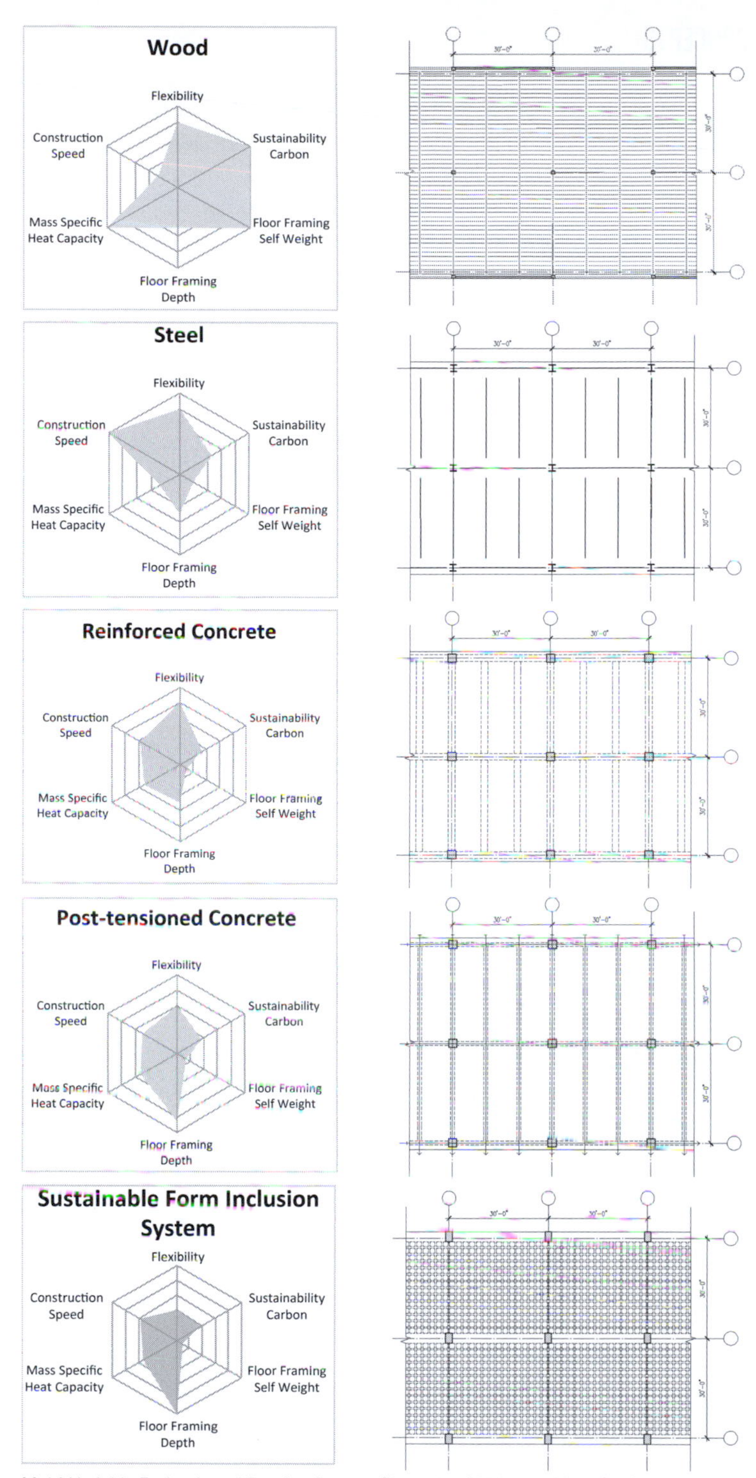

Multi-Variable Evaluation of Framing Systems [9 m × 9 m (30 ft × 30 ft) module]

2.6 CONCEPTS

Structural systems, from horizontal framing to vertical gravity and lateral load-resisting elements are important to define conceptually early in the design process. These systems directly respond to building use, availability of materials, construction time, and environmental impacts.

Integration of systems is also important for developing initial ideas of mechanical and exterior wall systems. Floor-to-floor heights are typically critical to the development of the initial ideas, cost, and project proforma. Typically, projects have a height limit; therefore, it is important to limit floor-to-floor heights while maximizing floor-to-ceiling heights, allowing spaces with greater daylight and flexibility.

The estimation of material quantities is important to establishing project costs. Therefore, a comparison of structural systems and materials required for each major component in the structure is important. Finally, prefabrication is an important consideration in reducing site construction time. Therefore, systems that lead to preassembled parts are an important consideration for concept development.

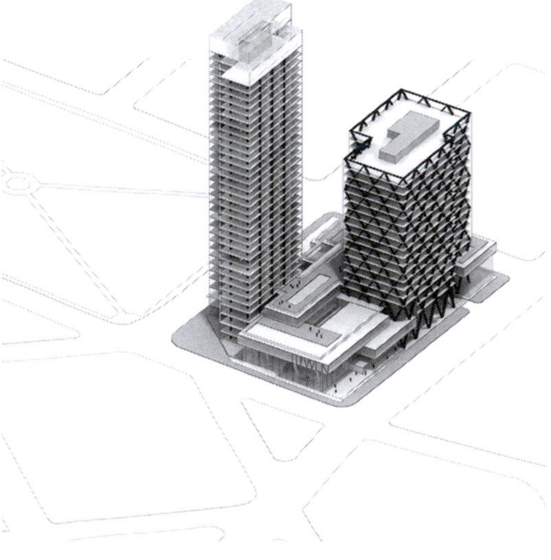

Vulcan Block 48 Design Competition, Seattle, WA

Structural Axonometric of Towers, Vulcan Block 48, Seattle, WA

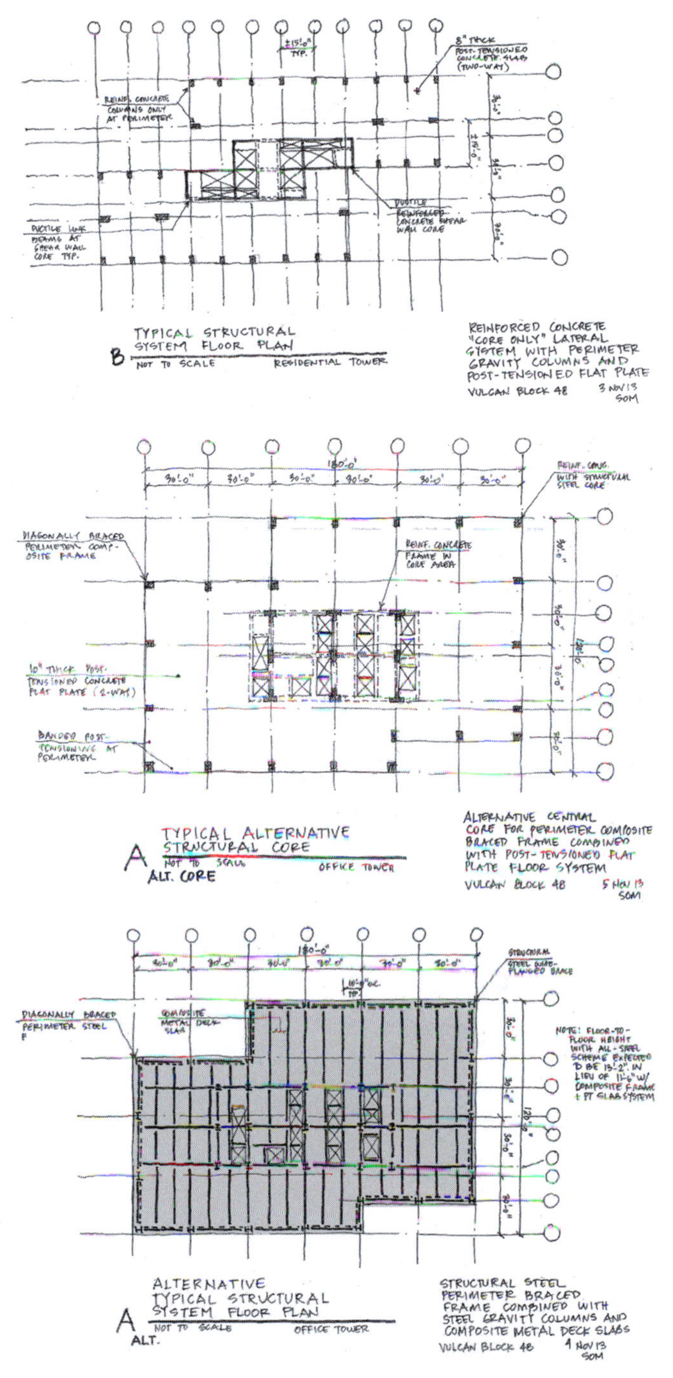

TOP TO BOTTOM

Reinforced Concrete Core-Only Structural System for Residential Tower, Vulcan Block 48, Seattle, WA

Composite Structural System for Office Tower, Vulcan Block 48, Seattle, WA

Steel Structural System for Office Tower, Vulcan Block 48, Seattle, WA

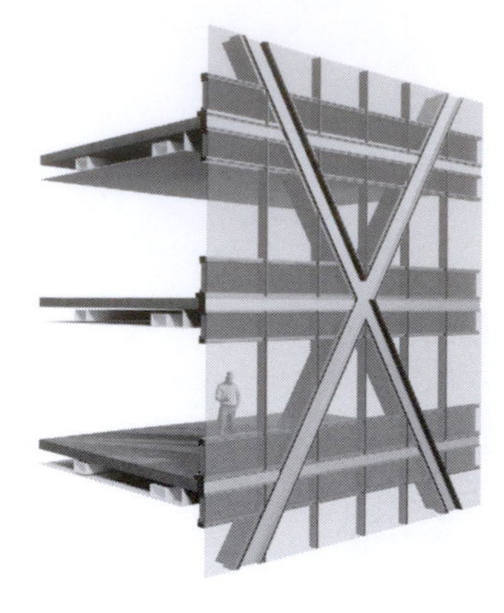

Office Tower Façade Concept, Vulcan Block 48, Seattle, WA

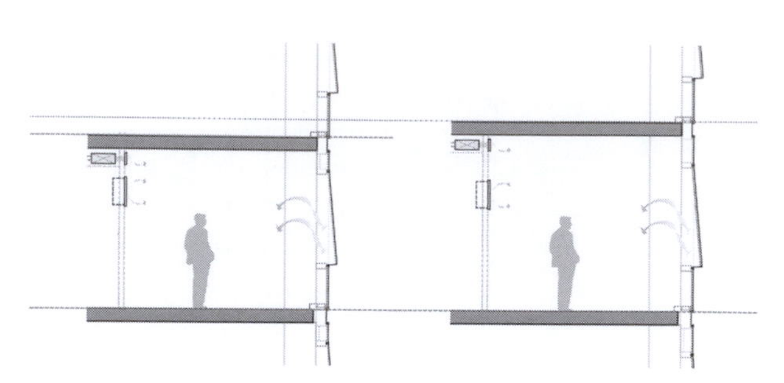

Typical Floor-to-Floor Section, Residential Tower, Vulcan Block 48, Seattle, WA

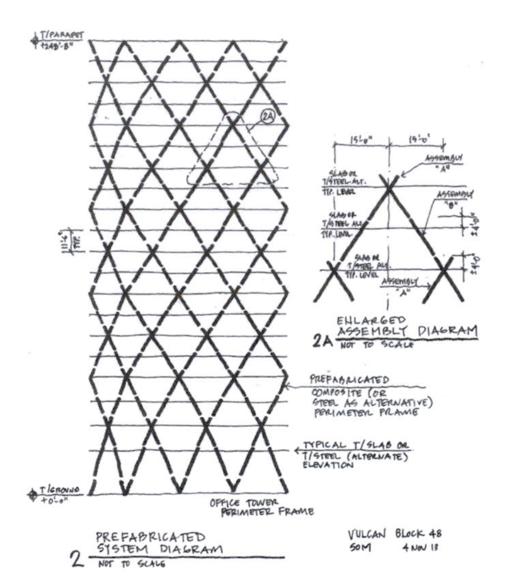

Prefabricated Perimeter Bracing System Concept, Office Tower, Vulcan Block 48, Seattle, WA

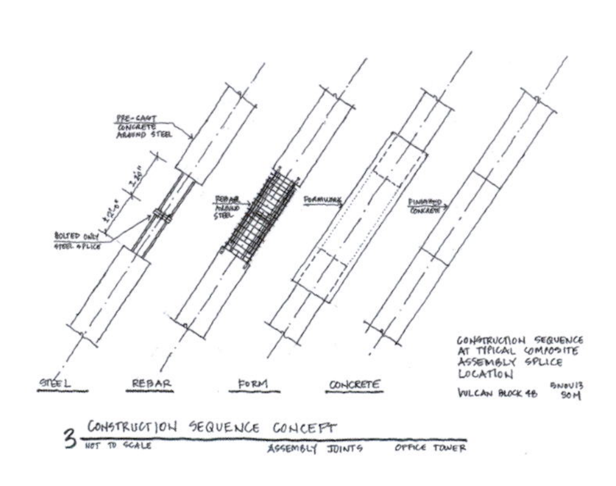

Construction Sequence Concept for Bracing System, Office Tower, Vulcan Block 48, Seattle, WA

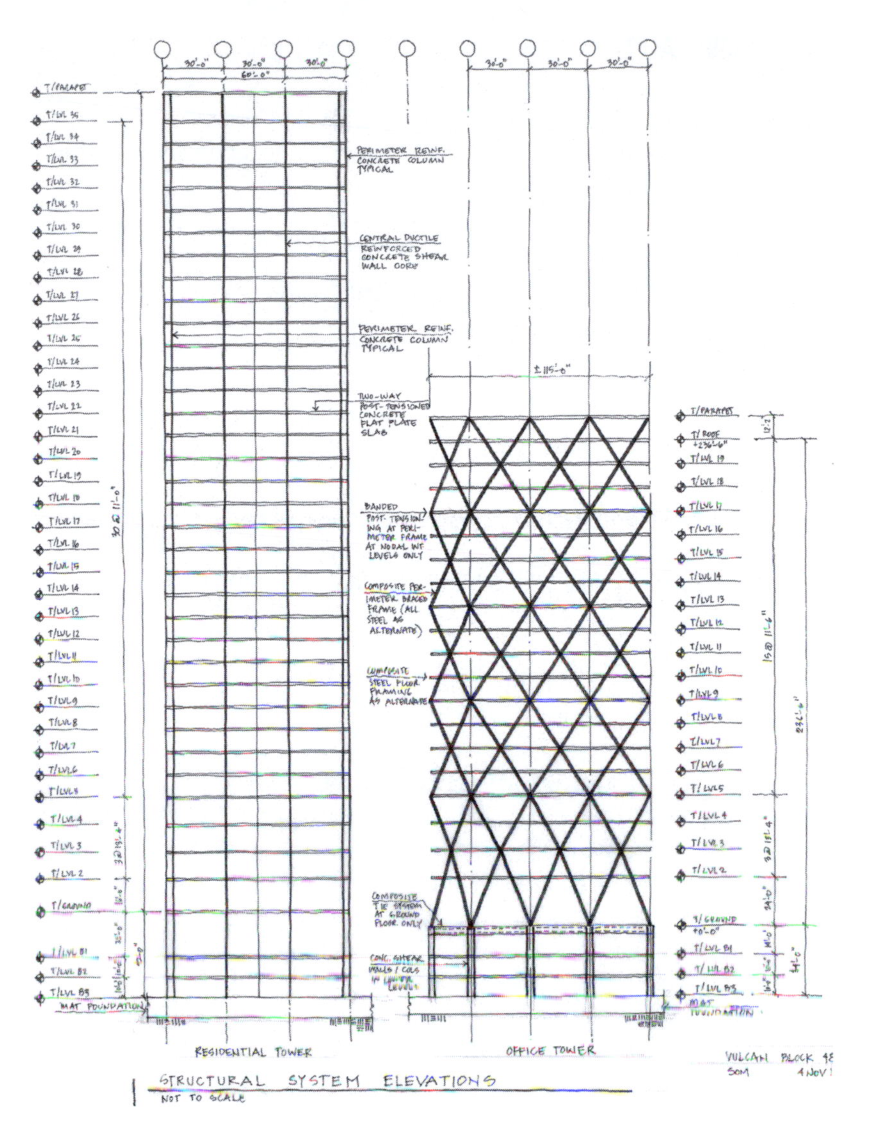

Structural System Elevations for Residential and Office Towers, Vulcan Block 48, Seattle, WA

2.7 HOW TALL?

It depends. Material strength is a limitation, but perhaps more important is the economic viability of the ultra-tall building. The economic viability is directly related to the net usable space that can be sold or leased. As towers increase in height, net usable area typically decreases.

Structural elements constructed vertically considering their own self-weight can be built to the following heights:

Concrete [compressive strength = 34.5 MPa (5000 psi or 720,000 psf)] = 1502 m (4965 ft)
Concrete [compressive strength = 69 MPa (10,000 psi or 1,440,000 psf)] = 3028 m (9931 ft)
Steel [yield strength = 345 MPa (50,000 psi or 7,200,000 psf)] = 4480 m (14,694 ft)
Steel [yield strength = 449 MPa (65,000 psi or 9,360,000 psf)] = 5824 m (19,102 ft)

The heights are determined by the following relationships:

Height = material strength (psf)/unit density (pcf)

where,

unit density of concrete = 2323 kg/m^3 (145 pcf)
unit density of steel = 7850 kg/m^3 (490 pcf)

Researchers and economists have concluded that a minimum net floor area ratio (NFA) of 75% is typically required to make a tall building profitable. Lower NFA values are common, many between 70–75% as documented for tall buildings constructed through the 1990s. Recently, developers have demanded NFA ratios of 80%, up to even 90%. These targets are increasingly challenging since the average height of newly constructed tall buildings continues to increase with proportional demand on building systems, and consequently, sizes of these systems. When building heights become significant (height > 200 m (656 ft)), NFA efficiencies greater than 75% are even more difficult to achieve.

On average, the gross floor area (GFA) consumed by building systems is 23%. Included are 12% core program, 5% structural area, 4% elevator shaft area, 1% mechanical, electrical, plumbing (MEP) shaft area, and 1% stair area. Core program consists of corridors, vestibules, lobbies, electrical and plumbing closets, janitorial closets, etc. The structural area is the plan extent of structural systems, including enclosing finishes.

As building heights increase, so does the space required for vertical transportation, mechanical systems, and structure. For the purpose of understanding the impact of these systems a set of assumptions are considered in the following sections.

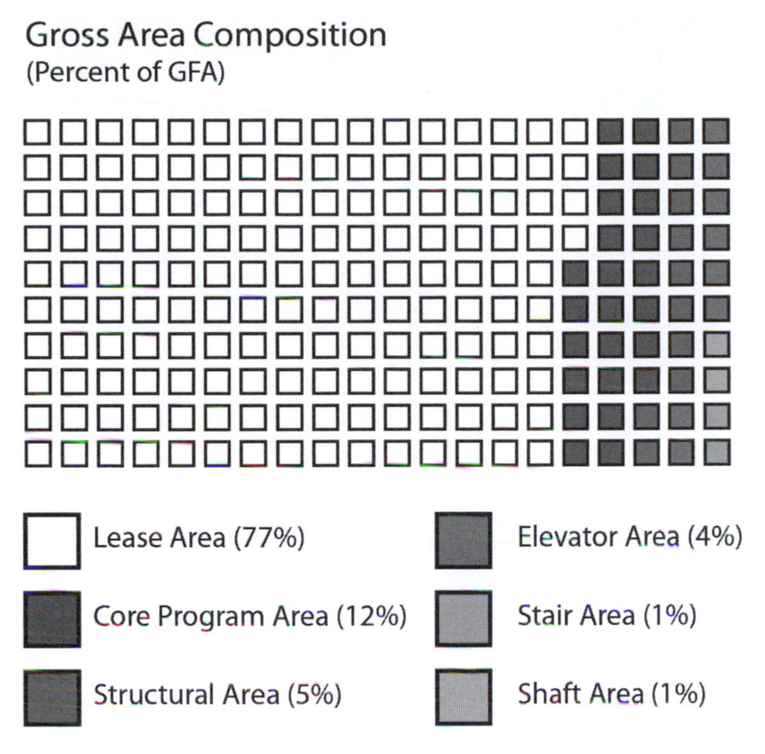

Average composition of GFA Resulting from Survey of Constructed Buildings

Structural Systems

Floor area required for structural elements such as columns, walls, and braces are estimated considering a self-weight of the structure based on material quantity estimation methods employed by the EA Tool™, assumed superimposed dead load of 0.7 kPa (15 psf), and live load of 3.8 kPa (80 psf). These load estimations are applied uniformly over the gross floor area and the total gravity weight is summed from top of building to base. This total load at each floor is divided by the selected material yield strength. To account for additional material corresponding to the lateral force-resisting system, a factor is applied to the yield strength. For high seismic regions a factor of 0.25 is applied to the specified yield strength, whereas a factor of 0.4 is used for high wind. When wind or seismic is considered moderate, a factor of 0.5 is used. In all cases, a minimum structural area of 3% of the total floor area is used.

Through this process, a required plan area of structural material is determined considering the building's form, height, material, and applied gravity and lateral loads. For steel, the plan extent of material is relatively small, but often steel shapes must be fireproofed and enclosed in finishes. As such, calculated structural steel floor area is multiplied by 10 to account for fireproofing and rectangular enclosure finishes.

Elevator Systems

Typically, a single cab elevator requires 9 m² (96.8 ft²) floor area. A tower under 45 stories will often have six to eight passenger elevators depending on the use. For buildings above 45 stories, more extensive groups of elevators, up to 18, can occur at a single floor. In very tall buildings, elevator groups will stack and sky lobbies will be introduced every 45 floors. Groups of six elevators can serve approximately 15 floors each. If a group of 18 elevators occurred in a 45-story module of a tall tower, three groups of six passenger elevators would service 15 floors each. The elevators which service the lower 15-floor sections would stop at the top of their respective zones and that floor area would be utilized for increased NFA. Allowances for one service elevator and one sky lobby elevator per sky lobby are included.

Building Service Systems

Allowances are utilized for average core program area of 20%. This average space will vary slightly with height.

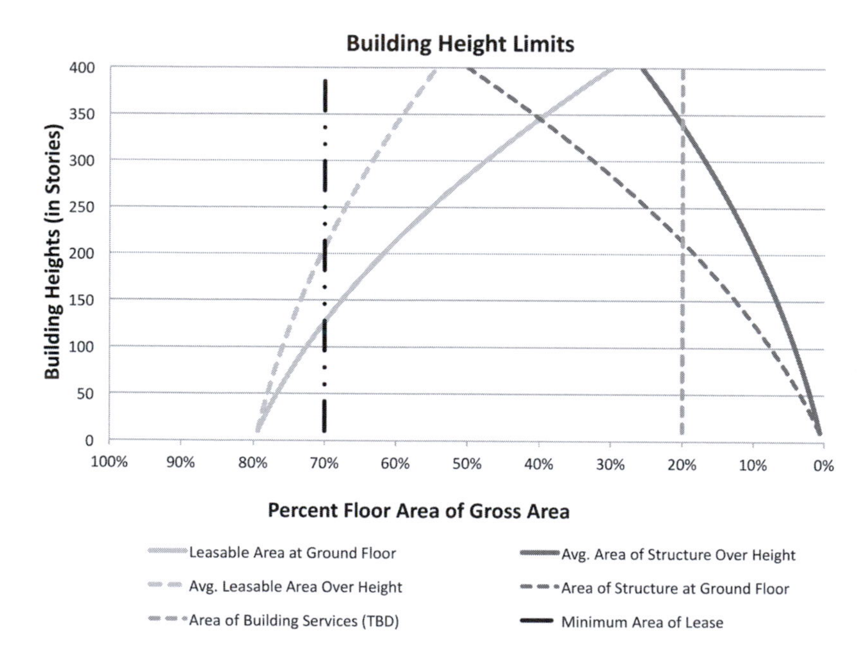

Relationship of Building Height to Floor Area

Based on the relationship, the average structural area increases from a theoretical zero at the base to about 25% at 400 stories. The area of structure at the ground floor varies from a theoretical zero to approximately 50%, if the structure was 400 stories tall. A similar consideration is made for occupied area averaged over the height of the building and at the ground level.

CHAPTER 3
SITE

THE PRIMARY SITE CONSIDERATIONS for tower design include the effects of wind, seismic, and geotechnical conditions. The conditions may be code-defined or derived from specific site conditions. The site conditions can be modeled analytically to replicate expected behavior during expected events.

Structures 200 m (656 ft) or more in height, even those consisting of reinforced concrete (which has greater mass than structural steel) and located in moderate to high seismic areas, are usually controlled by wind effects rather than seismicity. This by no means relaxes the required ductility, detailing, and redundancy for the structure, but it does mean that the structure is flexible with a significantly long fundamental period of approximately 5 seconds or longer, attracting less inertial forces than a shorter structure with a shorter period.

Poor soil conditions, near-fault effects, and potential earthquake intensity must be considered and may change the governing behavior. In fact, certain critical elements within the superstructure may be considered to perform elastically in even a rare earthquake event (475-year event, 10% probability of exceedance in 50 years). For instance, steel members located within an outrigger truss system that are intermittently located within the tower may require considerations for this level of force to achieve satisfactory performance even in an extreme seismic event.

3.1 WIND

3.1.1 General Effects

Direct positive pressure is exerted on the surface facing (windward faces) or perpendicular to the wind. This phenomenon is directly impacted by the moving air mass and generally produces the greatest force on the structure

FACING PAGE
Burj Khalifa Foundations,
Dubai, UAE

Wind Flow, John Hancock Center, Chicago, IL

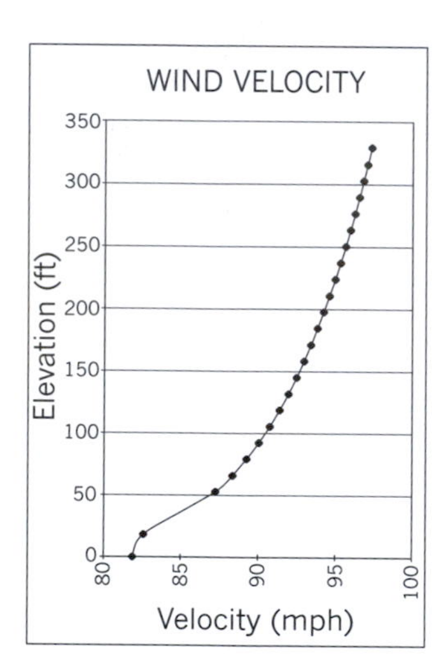

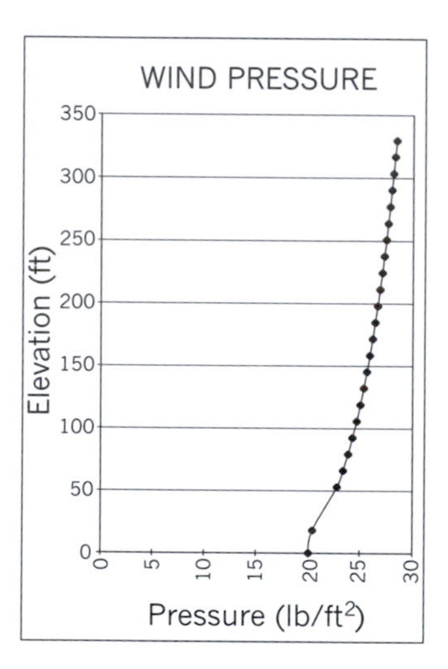

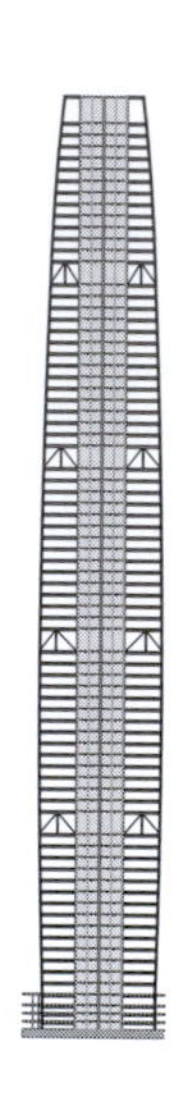

Comparison of Wind Velocity and Wind Pressure

unless the tower is highly streamlined in form. Negative pressure or suction typically occurs on the leeward (opposite face from the wind) side of the tower. Since the winds flow like a liquid, there are drag effects on the surfaces parallel to the direction of the wind. These surfaces may also have positive or negative pressures on them, but it is the drag effect that adds to the general force on the tower. The combination of these three effects generally results in the net force on a tower. However, for very tall or slender structures their dynamic characteristics can produce amplified forces. Across-wind or lift motion is common for these structures. In fact, many of these taller structures are controlled by this behavior. This dynamic effect could exist at even low velocities if the velocity of the wind causes force pulses through vortex shedding that match the natural period of vibration of the structure.

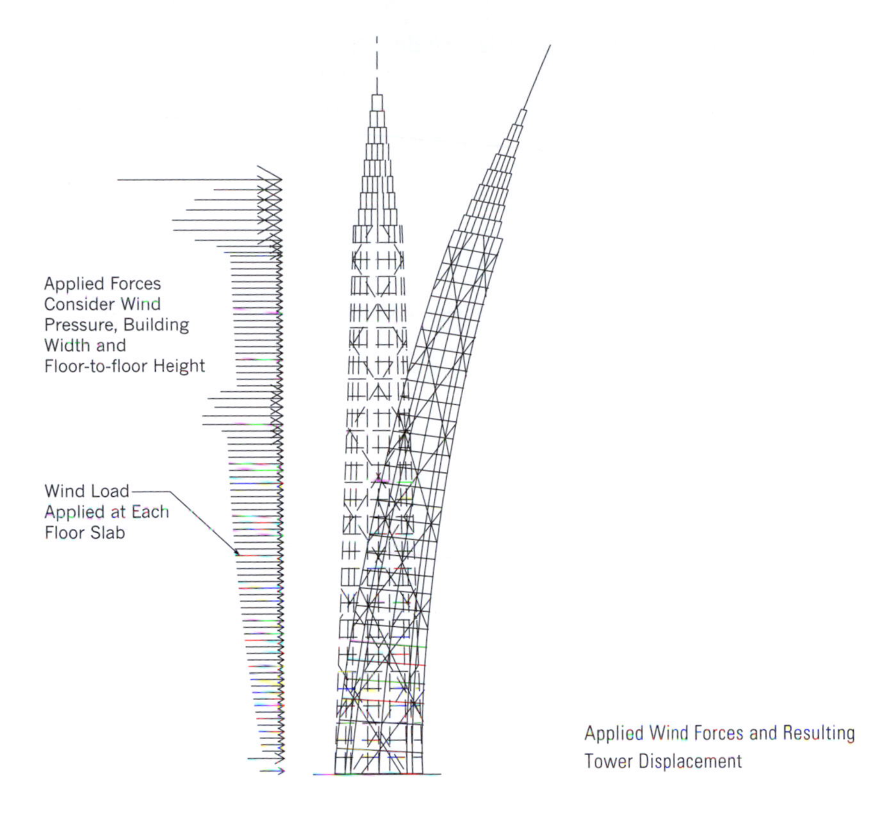

Applied Wind Forces and Resulting Tower Displacement

Generally, the relationship between wind velocity and pressure is:

$$P = 0.003V^2$$

where:

P = equivalent static pressure on a stationary object (lbs/ft^2)
V = wind velocity (miles/hr)

Although wind conditions are generalized for a given geographic area, the local terrain at a site has a substantial effect on the pressures expected. For instance, the applied pressures expected for a structure located in open terrain are significantly higher than those expected in an urban setting where tall structures may surround the site.

Code-defined wind criteria must be used as the basis for all tall building design; however, these criteria are generally too conservative for the tall tower. Buildings 40 stories or taller should be considered for wind tunnel studies that evaluate realistic structural behavior. These studies result in a rational evaluation of the in-situ wind climate and usually lower base building design forces and provide accurate, local wind effects on cladding and on pedestrians at the ground plane.

Wind tunnel studies should include:

a. proximity modeling/wind climate (detailed modeling of structures within 0.8 km (0.5 miles) of site) and wind environment analysis based on historic data
b. pressure tap modeling of exterior walls
c. pedestrian wind analysis
d. force–balance structural modeling
e. aero-elastic structural modeling (consider for heights over 300 m (984 ft))

3.1.2 Code Requirements

The American Society of Civil Engineers Minimum Design Loads for Buildings and Other Structures (ASCE 7-10), Chapter 31 permits rational wind tunnel studies to determine loads on any building or structure in lieu of code formulas; however, many governing jurisdictions require that minimum code-defined loads must be used for strength design. In most instances, in addition to exterior wall design, the rational wind tunnel studies can be used to evaluate the structure for serviceability, including drift and accelerations. This usually leads to a considerable reduction in base building stiffness. According to the 2012 International Building Code (IBC), which includes most of the requirements of ASCE 7-10, minimum base building structural wind load is determined with the following design procedure:

1. Determine the risk category of the building.
2. Determine the ultimate design wind speed V_{ult} and the wind directionality factor K_d.
3. Calculate the exposure category or exposure categories and velocity pressure exposure coefficient K_z or K_h.
4. Calculate the topographic factor K_{zt}.
5. Determine the gust effect factor G or G_f.
6. Determine the enclosure classification GC_{pi}.
7. Calculate the internal pressure coefficient GC_{pf}.
8. Calculate the external pressure coefficients C_p, C_N or GC_{pi}.
9. Calculate the velocity pressure q_z or q_h as applicable.
10. Calculate the design wind pressure p, where p for rigid buildings is:

$$p = qGC_p - q_1(GC_{pi})$$

and

$$q^z = 0.00256 K_z K_{zi} K_d V_{ult}^2 \text{ (lbs/ft}^2\text{)}$$

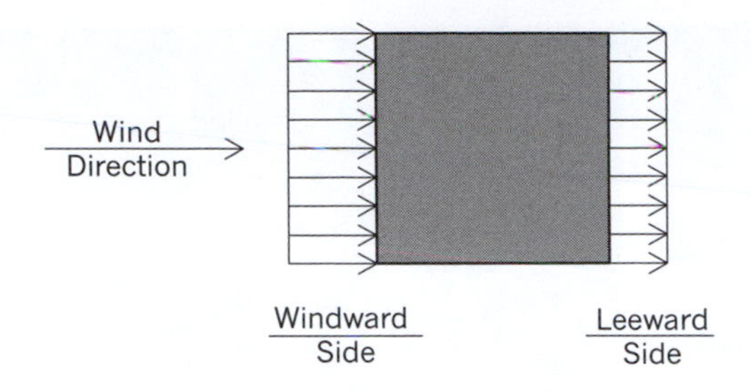

The IBC is based on wind velocities measured 10 m (33 ft) above the ground and based on 3-second gusts. For the IBC 2012, the wind speed maps specifying these velocities were updated to display the ultimate design level speeds. To convert to previous code editions' wind speed values, which may still be used for specific serviceability checks, the following equation is provided:

$$V_{asd} = V_{ult}\sqrt{0.6}$$

where:

V_{ult} = 3-second gust ultimate design wind speed (available from wind speed maps in the IBC 2012)

V_{asd} = 3-second gust nominal design wind speed (equivalent to the basic wind speed from wind speed maps in IBC editions prior to 2012)

According to the 1997 Uniform Building Code (UBC), base building design wind pressure is defined as:

$$P = C_e C_q q_s I_w$$

where:

P = design wind pressure
C_e = combined height, exposure and gust factor coefficient
C_q = pressure coefficient for the structure
q_s = wind stagnation pressure at the standard height of 10 meters (33 feet) based on the fastest mile wind speed (highest average wind speed based on the time required for a mile-long sample of air to pass a fixed point)
I_w = importance factor

Most building codes are based on a 50-year return wind event for strength and serviceability design of building structures. There are cases, however, where a 100-year return wind must be considered for design particularly related to building drift. This increase in design pressure is usually at least 10%.

Jin Mao Tower Tunnel Wind Tunnel Testing

3.1.3 Rational Wind Considerations

The rational wind can be considered in two components: static and dynamic. Magnitude, direction, and proximity to neighboring structures (both current and future) are important. Frequently, future planned buildings (if known to be part of a master plan at the time of design) could dynamically excite the structure, causing magnification of force levels and a more conservative, yet appropriate design. For instance, in the case of the Jin Mao Building in Shanghai, two taller towers were planned within neighboring city blocks resulting in design forces that were controlled by dynamic effects from vortex shedding of wind from the neighboring structures. This behavior magnified the forces on Jin Mao by 33%.

Computational Fluid Dynamic (CFD) modeling techniques have been used more frequently in recent years to understand the rational wind behavior of tall buildings. Although an excellent method for determining general wind effects, physical modeling in a wind tunnel environment usually leads to most accurate results.

San Andreas Fault, Carrizo Plain, CA, Building Damage as a Result of the 1995 Kobe Earthquake

3.2 SEISMICITY

3.2.1 Intensity

The intensity of an earthquake is based on a qualitative assessment of damage and other observed effects on people, buildings, and other features. Intensity varies based on location within an affected region. An earthquake in a densely populated area may result in many deaths and considerable damage where the same earthquake may result in no damage or deaths in remote areas. The scale used most to evaluate the subjective intensity is the Modified Mercalli Intensity Scale (MMI) developed in 1931 by American seismologists Harry Wood and Frank Neumann. The scale consists of 12 increasing levels of intensity expressed in Roman numerals. The scale ranges from imperceptible motion (Intensity I) to catastrophic destruction (Intensity XII). A qualitative description of the complete scale is as follows:

Intensity I — Not felt except by very few under especially favorable conditions.

Intensity II — Felt only by a few persons at rest, especially by those on upper floors of buildings. Delicately suspended objects may swing.

Intensity III — Felt quite noticeably by persons indoors, especially in upper floors of buildings. Many people do not recognize it as an earthquake. Standing vehicles may rock slightly. Vibrations similar to the passing of a truck. Duration estimated.

Results of Strong Ground Motion, Olive View Hospital (1971), Sylmar, CA

Intensity IV During the day, felt indoors by many, outdoors by a few. At night, some awakened. Dishes, windows, doors disturbed; walls make cracking sound. Sensation like heavy truck striking a building. Standing vehicles rock noticeably.

Intensity V Felt by nearly everyone; many awakened. Some dishes, windows broken. Unstable objects overturned. Pendulum clocks may stop.

Intensity VI Felt by all, many frightened. Some furniture moved. A few instances of fallen plaster. Damage slight.

Intensity VII Damage negligible in buildings of good design and construction; slight to moderate in well-built ordinary structures; considerable damage in poorly built structures. Some chimneys broken.

Intensity VIII Damage slight in specially designed structures; considerable damage in ordinary substantial buildings, with partial collapse. Damage great in poorly built structures. Fallen chimneys, factory stacks, columns, monuments, walls. Heavy furniture overturned.

Intensity IX Damage considerable in specially designed structures; well-designed frame structures thrown out of plumb. Damage great in substantial buildings, with partial collapse. Buildings shifted off foundations.

Intensity X Some well-built wooden structures destroyed; most masonry and frame structure with foundations destroyed. Rails bent greatly.

Intensity XI Few, if any, masonry structures remain standing. Bridges destroyed. Rails bent greatly.

Intensity XII Damage total. Lines of sight and level are destroyed. Objects thrown into the air.

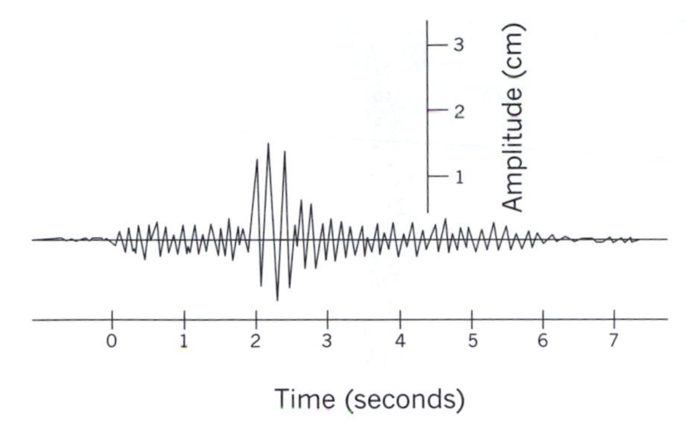

Seismograph Trace

3.2.2 Magnitude

The most commonly used measure of an earthquake's strength is determined from a scale developed by Charles F. Richter of the California Institute of Technology in 1935. The magnitude, M, of an earthquake is determined from the base ten logarithm of the maximum amplitude of oscillation measured by a seismograph.

$$M = \log_{10}(A/A_o)$$

where,

A = measured maximum amplitude

A_o = measured amplitude of an earthquake of standard size (calibration earthquake)

A_o generally equal to 3.94×10^{-5} in (0.001mm).

The above equation assumes that the seismograph and the epicenter are 100 km (62 miles) apart. For other distances a nomograph must be used to calculate M.

Since the equation used to calculate M is based on a logarithmic scale, each whole number increase in magnitude represents a ten-fold increase in measured amplitude. The Richter magnitude M is typically expressed in whole and decimal numbers. For example, 5.3 generally corresponds to a moderate earthquake, 7.3 generally corresponds to a strong earthquake, and 7.5 or above corresponds to a great earthquake. Earthquakes of magnitude 2.0 or less are known as microearthquakes and occur daily in the San Francisco Bay Area. The 1989 Loma Prieta Earthquake measured 7.1 on the Richter Scale with the 1906 San Francisco Earthquake corresponding to 8.3. The largest recorded earthquake was the great Chilean Earthquake of 1960 where a magnitude of 9.5 was recorded.

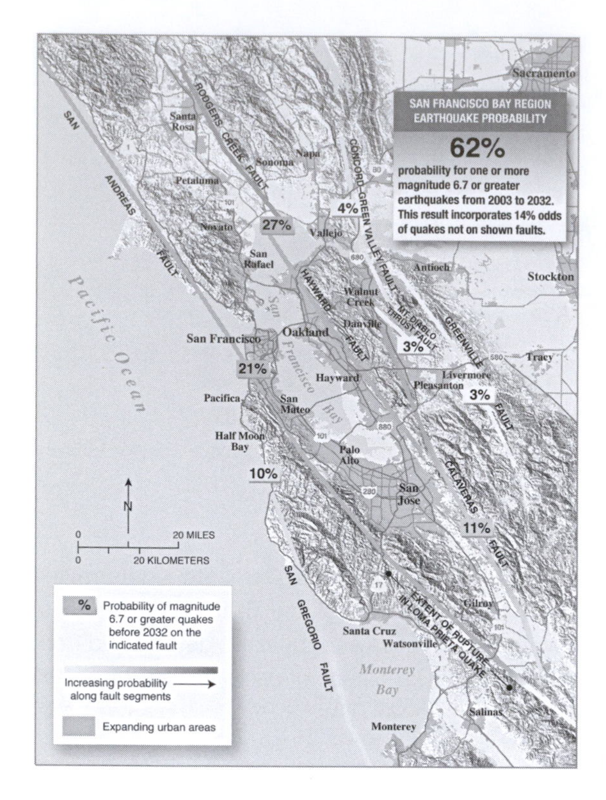

Probability of a 6.7 Magnitude Earthquake or Greater in San Francisco before Year 2032

The seismometer is the detecting and recording portion of a larger apparatus known as a seismograph. Seismometers are pendulum-type devices that are mounted on the ground and measure the displacement of the ground with respect to a stationary reference point. Since the device can record only one orthogonal direction, three seismometers are required to record all components of ground motion (two translational, one vertical). The major movement during an earthquake occurs during the strong phase. The longer the earthquake shakes, the more energy is absorbed by the buildings resulting in increased damage based on duration. The 1940 El Centro Earthquake (magnitude 7.1) had 10 seconds of strong ground motion and the 1989 Loma Prieta Earthquake (magnitude 7.1) lasted only 10–15 seconds. By contrast the 1985 Chilean Earthquake (magnitude 7.8) lasted 80 seconds and the 1985 Mexico City Earthquake (magnitude 8.1) lasted 60 seconds. There is debate in California, with no consensus, on whether long duration earthquakes can occur. The assumed strong ground shaking for the current 1997 UBC Zone 4 earthquake is 10 to 20 seconds.

Soil Liquefaction (1995), Kobe, Japan

3.2.3 Energy

The energy released during a seismic event can be correlated to the earthquake's magnitude. In 1956, Beno Gutenberg and Richter determined an approximate correlation where the radiated energy (ergs) is less than the total energy released, with the difference related to indeterminate heat and other non-elastic effects.

$$\log_{10} E = 11.8 + 1.5M$$

Since the relationship between magnitude and energy is logarithmic with the associated factors, an earthquake of magnitude 6 radiates approximately 32 times the energy of magnitude 5. In other words, it would take 32 smaller earthquakes to release the same energy as of one earthquake one magnitude larger.

3.2.4 Peak Ground Acceleration

Peak ground or maximum acceleration (PGA) is measured by an accelerometer and is an important characteristic of an earthquake oscillatory response. This value is frequently expressed in terms of a fraction or percent of gravitational acceleration. If a building structure had infinitely high stiffness (period of vibration of essentially zero), the structure would move with the ground with no relative displacement of building mass to its foundation. For instance, the peak ground acceleration measured during the 1971 San Fernando Earthquake was 1.25g or 125%g or 12.3 m/s^2 (40.3 ft/s^2). The peak ground acceleration measured during the Loma Prieta Earthquake was 0.65g or 65%g or 6.38 m/s^2 (20.9 ft/s^2).

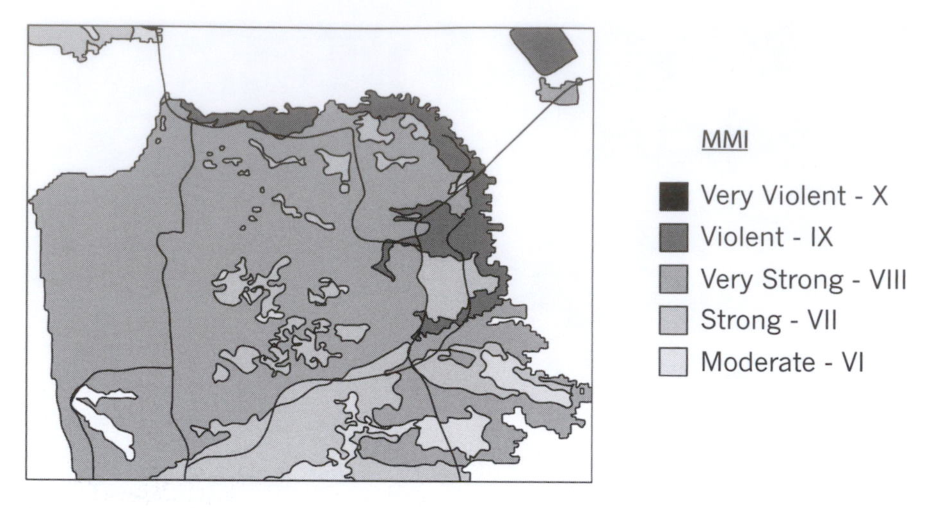

Projected Damage (MMI) in San Francisco from 1989 Loma Prieta Earthquake (M=7.1)

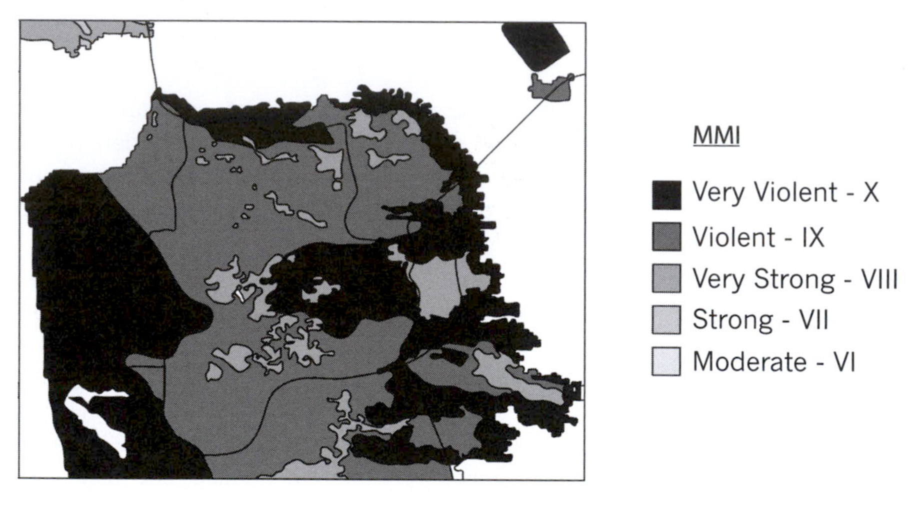

Projected Damage (MMI) in San Francisco from 1906 Earthquake (M=8.3)

3.2.5 Correlation of Intensity, Magnitude, and Peak Ground Acceleration

No exact correlation between intensity, magnitude, and peak ground acceleration exists since damage is dependent on many factors, including earthquake duration and the way that the structure was designed and constructed. For instance, buildings designed in remote locations of developing countries will likely perform much worse than structures designed in major urban areas of developed countries. However, within a geographic region with common design and construction practices, a fairly good correlation may be drawn between intensity, magnitude, and peak ground acceleration.

MMI	PGA	Approx. Magnitude
IV	0.03g and below	
V	0.03g–0.08g	5.0
VI	0.08g–0.15g	5.5
VII	0.15g–0.25g	6.0
VIII	0.25g–0.45g	6.5–7.5
IX	0.45g–0.60g	8.0
X	0.60g–0.80g	8.5
XI	0.80–0.90g	
XII	0.90g and above	

In addition, seismic zones as defined by the 1997 UBC can be correlated to an earthquake's magnitude and peak ground acceleration.

Seismic Zone	PGA	Max. Magnitude
0	0.04g	4.3
1	0.075g	4.7
2A	0.15g	5.5
2B	0.20g	5.9
3	0.30g	6.6
4	0.40g	7.2

In certain limited areas or micro-zones, peak ground accelerations may vary significantly. This variance is primarily attributed to local site soil conditions. During the Loma Prieta Earthquake, peak accelerations measured in San Francisco were generally not greater than 0.09g, but peak accelerations recorded at the Bay Bridge, Golden Gate Bridge, and the San Francisco Airport were 0.22–0.33g, 0.24g, and 0.33g respectively. After the Mexico City Earthquake of 1985, micro-zones were incorporated into the rebuilding plan.

3.2.6 Earthquake, Site, and Building Period

Earthquakes release energy in different frequency ranges. The period (or the natural frequency) of a vibration, the time it takes for one full oscillatory cycle, is the characteristic of motion that affects sites and structures. If the site (soil) has a natural frequency of vibration that corresponds to the predominant earthquake frequency, site movement can be greatly amplified through a phenomenon called resonance. Structures located on these sites can experience amplified forces. Soil characteristics such as density, bearing strength, moisture content, compressibility, and tendency to liquefy all may affect the site period.

Theoretically, a structure with zero damping, when displaced laterally by an earthquake, will oscillate back and forth indefinitely with a regular period. As damping is introduced, the motion will eventually stop. The building period is not the site period; however, if these periods are close to one another, resonance could occur with a large magnification of forces that the structure must resist.

3.2.7 Probability of Exceedance and Return Period

Earthquakes are commonly described by the percent probability of being exceeded in a defined number of years. For instance, a code-defined design basis earthquake is typically referred to as having a 10% probability of being exceeded in 50 years. Another way of describing this earthquake design level is through "return period." For this code-defined earthquake (10% probability of exceedance in 50 years) the earthquake is also known as having a 475-year return period or sometimes referred to as a 475-year event. The following describes the conversion between return period and probability of exceedance.

$$RP = T / r^*$$

where:

$r^* = r\,(1 + 0.5r)$
RP = return period
t = target year of exceedance
r = % probability of exceedance

Therefore, for a 10% probability of exceedance in 50 years:

$$RP = 50 / 0.10\,(1{+}0.5\,(0.1)) = 476.2 \approx 475$$

The table below includes commonly used probability of exceedances and return periods.

3.2.8 Spectral Acceleration

Measured amplitude of an earthquake over time during a seismic event is not regular. It is difficult to determine how a structure behaves at all times during an earthquake consisting of random pulses. In many cases it is not necessary to evaluate the entire time history response of the structure because the structure is likely more affected by the peak acceleration rather than smaller accelerations that occur during the earthquake. The spectral acceleration is the cumulative result of the interaction of the structure's dynamic characteristics with the specific energy content of an earthquake.

Probability of Exceedance / Return Period Table					
Event	**r**	**T**	**r***	**RP**	**RP (Rounded)**
63% in 50 Years	0.63	50	0.1315	60.4	60
10% in 50 Years	0.10	50	0.105	476.2	475
5% in 50 Years	0.05	50	0.05125	975.6	975
2% in 50 Years	0.02	50	0.202	2475.2	2475
10% in 100 Years	0.10	100	0.105	952.4	975

The spectral acceleration is the maximum acceleration experienced by a single degree of freedom vibratory system of a given period (period of the structure) in a given earthquake. The maximum velocity and displacement are known as the spectral velocity and spectral displacement respectively.

The maximum building acceleration is typically higher than the peak ground acceleration; therefore, these values should not be confused. The ratio of the building to peak ground acceleration depends on the building period. For an infinitely stiff structure (period = 0 sec) the ratio is 1.0. For short period structures in California considering a 5% damped building and a hazard level (probability of occurrence) of 10% in a 50-year period located on rock or other firm soil, the ratio is approximately 2.0 to 2.5 times the peak ground acceleration (spectral amplification).

Response spectra commonly used in design are developed based on spectral accelerations. These spectra may be site specific or code-defined.

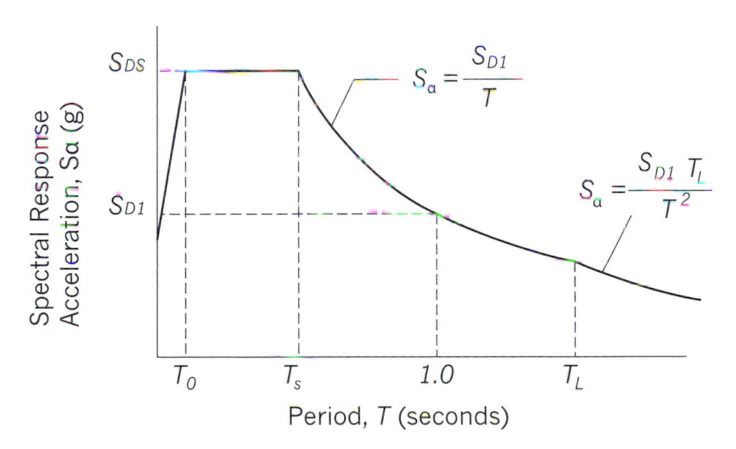

2012 International Building Code Seismic Response Spectrum

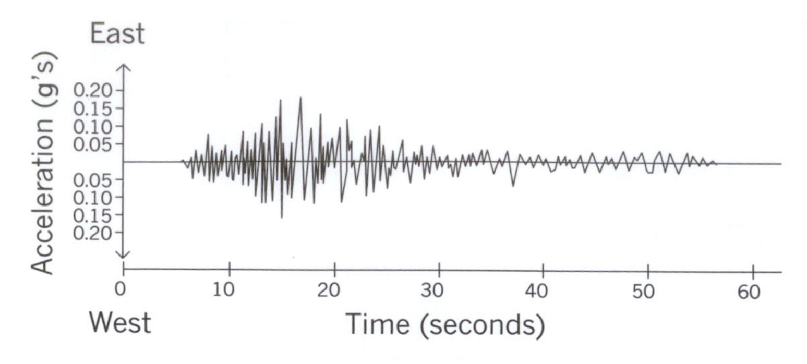

Seismic Time History Record

3.2.9 Design and Maximum Considered Earthquakes

The design basis level earthquake as recognized by the 1997 UBC and current codes is based on an earthquake that has a 10% probability of being exceeded in 50 years (approximately a 475-year event). This design level is based on the reasonable likelihood that an earthquake of this magnitude will occur during the life of the structure. At this level of seismicity, the structure is expected to be damaged, but not collapse, and life safety protected.

The maximum considered earthquake represents the maximum earthquake expected at a site. Generally this is an event that has a 2% probability of being exceeded in 50 years (approximately a 2475-year event). Typically structures are designed for stability (collapse prevention) in this earthquake event but higher performance goals (i.e., life safety) may be required for important/essential facilities such as hospitals or police stations for this extreme event.

3.2.10 Levels of Seismic Performance

Operational Level (O): Backup utility services maintain functions; very little damage.

Immediate Occupancy Level (IO): The building receives a "green tag" (safe to occupy) inspection rating; any repairs are minor.

Life Safety Level (LS): Structure remains stable and has significant reserve capacity; hazardous nonstructural damage is controlled.

Collapse Prevention Level (CP): The building remains standing, but only barely; any other damage or loss is acceptable.

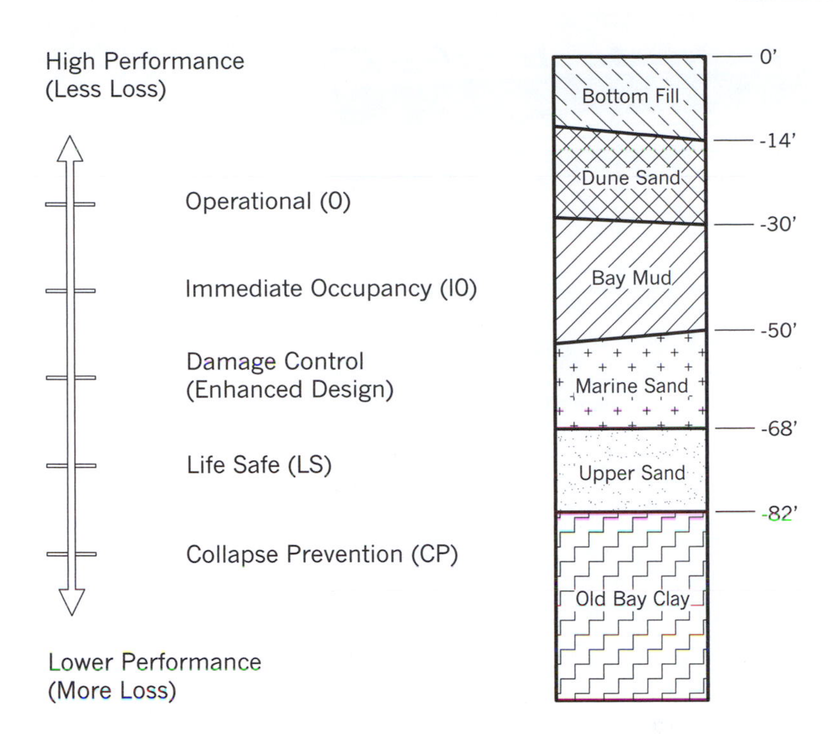

Considerations for Seismic Performance

Typical Soil Conditions from South of Market Street Sites in San Francisco

3.3 SOILS

Geotechnical conditions vary widely for sites of ultra-tall structures. Mechanics of the site soil conditions include stability, water effects, and anticipated deformations. Soil conditions may vary from bedrock to sand to clay, bedrock having the best geotechnical characteristics with dense sand having similar traits. Sand provides good foundation support since settlement is elastic (associated with initial loading from the structure), but could be difficult to accommodate during construction and could liquefy (complete loss of shear strength) when saturated and subjected to lateral seismic loads. Clay could provide excellent foundation support especially if pre-consolidated, but must be considered for both initial loading effects and long-term creep effects due to consolidation. Clay could prove to be excellent for site excavations.

Spread footings usually prove to be the most cost-effective foundation solution, followed by mat foundations. When bearing capacities are low or applied loads are high, deep foundations consisting of piles or caissons are usually required. The following is a general summary considering foundation type.

Mat Foundation Prior to Concrete Pour,
Burj Khalifa, Dubai, UAE

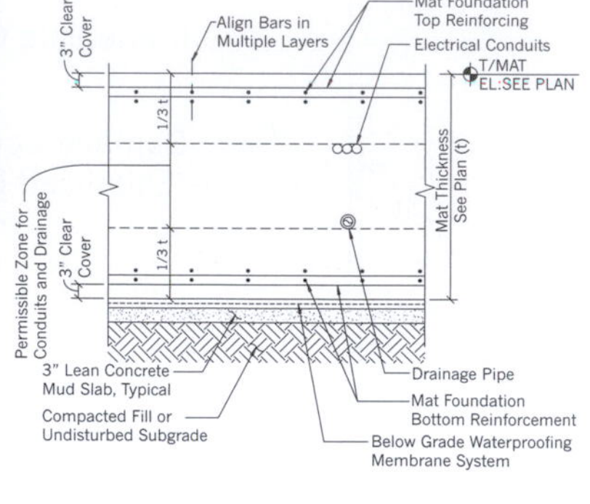

Mat Foundation Detail

3.3.1 Spread or Continuous Wall Footings

Spread or continuous footings are used under individual columns or walls in conditions where bearing capacity is adequate for applied load. This system may be used on a single stratum, firm layer over soft layer, or reasonably soft layer over a firm layer. Immediate, differential, and consolidation settlements must be checked.

3.3.2 Mat Foundations

Mat foundations are used in similar applications as spread or continuous wall footings where spread or wall footing cover over 50% of the building area. Use is appropriate for heavy column loads with the mat system usually reducing differential and total settlements. Immediate, differential, and consolidation settlements must be checked.

Pile Foundation, Jin Mao Tower, Shanghai, China

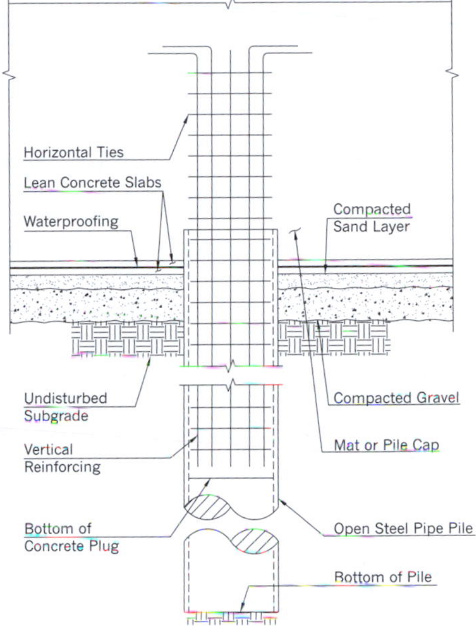

Pipe Pile Foundation Detail

3.3.3 Pile Foundations

Pile foundations are used in groups of two or more to support heavy column or wall loads. Reinforced concrete pile caps are used to transfer loads from columns or walls to the piles. Pile foundations provide an excellent solution for poor surface and near-surface soil conditions. This foundation system is a good solution for structures in areas for potential soil liquefaction. Piles are generally 20–50 m (65–164 ft) long below the lowest basement. Pile capacity is typically developed by skin friction, but end bearing may also be considered. Piles are usually designed to resist lateral loads (due to wind or seismic) in addition to vertical load. Bending on piles may be considered with heads fixed or pinned into pile caps. Piles typically consist of steel or concrete for tower structures (although timber could also be used). Corrosive soil conditions may require concrete (precast) to be used. H-piles in structural steel and 355 mm x 355 mm (14 in x 14 in) or 406 mm x 406 mm (16 in x 16 in) precast piles are common. Open steel pile piles have been used in conditions of dense sand and extremely high applied loads.

Caisson Construction, NBC Tower at Cityfront Center, Chicago, IL

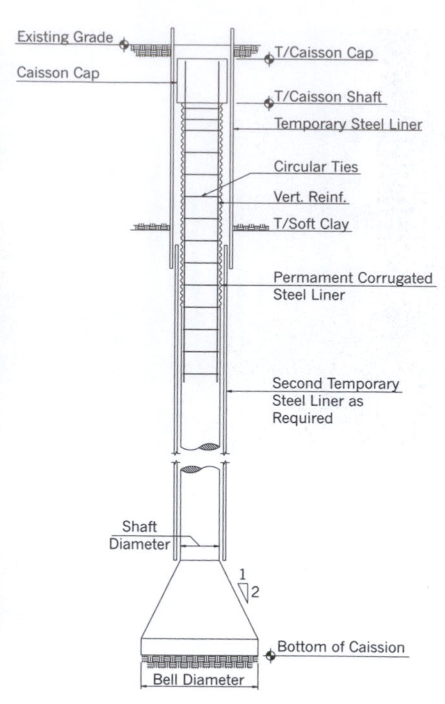

Caisson Foundation Detail

3.3.4 Caisson Foundations

Cast-in-place reinforced concrete caissons typically have a diameter of 750 mm (30 in) or more and may either be straight-shafted or belled. Bell diameters are typically three times the shaft diameter. Caisson foundations provide an excellent solution for poor surface and near-surface soil conditions. The capacity of this system is usually based on end bearing. End bearing of caissons is commonly founded in stiff clay (hardpan). Installation of caissons is very sensitive to soil conditions. Voids in shafts or bells are quite possible due to local soil instability during installation. Concrete may be placed under bentonite slurry to prevent soil instability during installation. The length of the caisson shaft usually varies from 8 to 50 m (26 to 164 ft). Caisson shaft design is typically based on the strength of concrete alone.

$$P_{cap} = A_c \times 0.25 f'_c$$

where:

P_{cap} = axial capacity of caisson
A_c = cross-sectional area of concrete
f^1_c = concrete strength

Slurry Wall Construction, Jin Mao Tower, Shanghai, China

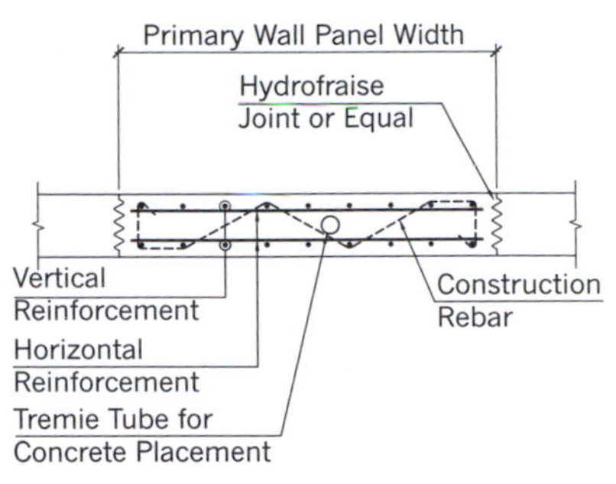

Slurry Wall Panel Detail

Slurry Wall Construction, Harvard University Northwest Science Building, Cambridge, MA

3.3.5 Basement/Foundation/Retaining Walls

Basement/foundation/retaining walls can be used in any soil condition, but usually require controlled, engineered backfill behind the walls. Where permanent water conditions exist, waterproofing is required. Slurry walls, cast under a bentonite slurry, provide temporary soil retention and permanent foundation walls. Bentonite caking at the exterior provides permanent waterproofing. Slurry walls are installed in panels, usually 4 m (15 ft) long, with shear keyways existing between panels. Reinforcing typically does not cross panel joints.

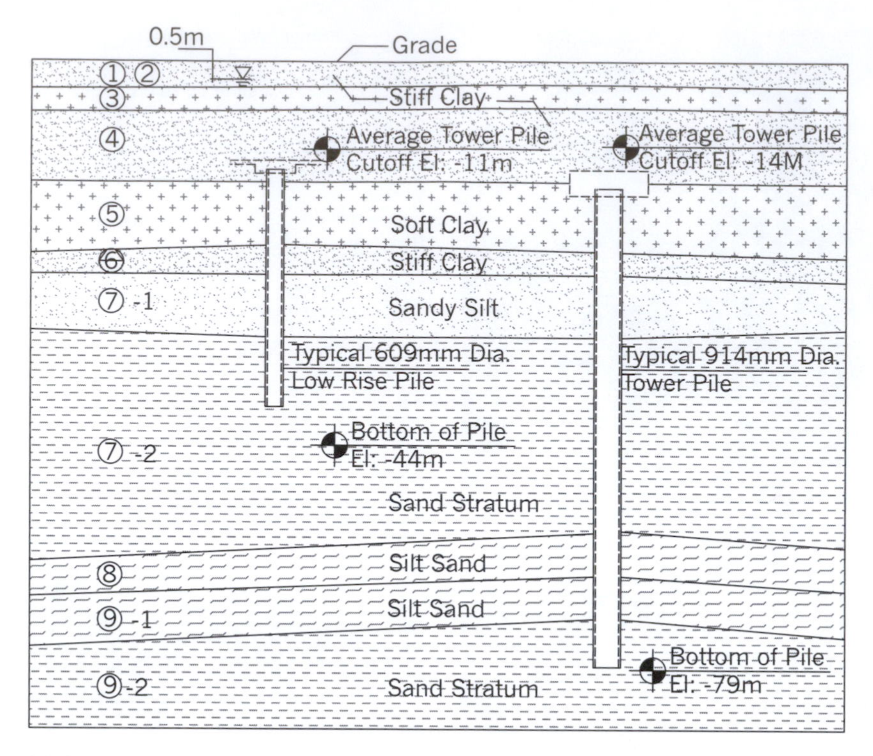

Soil Strata/Extent of Pile Foundation, Jin Mao Tower, Shanghai, China

3.3.6 Deep Foundation Considerations

Sites that do not have reachable bedrock can be considered for these structures; however, foundation systems become increasingly complex with both strength and settlement issues being critical. A bearing capacity of 480 kPa (10 ksf) usually represents a minimum threshold for design. A bearing capacity of 1900–2400 kPa (40–50 ksf) is more desirable. Pile or caisson foundations allow for adequate support where both skin friction and tip bearing can be used for the design. Piles or caissons should extend 3.0–4.5 m (10–15 ft) into bedrock through a top plain of weathered material that usually exists. Where bedrock does not exit, piles or caissons can be supported in deep stiff sands or hardpan clays. Care should be taken in establishing bearing elevations. Strength may be satisfied at certain soil layers, but these layers may exist over lower compressible layers that could cause adverse long-term settlement. Settlements of 75–125 mm (3–5 in) are not uncommon for pile supported (driven steel, precast concrete, or auger-cast concrete) ultra-tall structures. These settlements must be carefully considered for building entrance levels at grade or interfaces with neighboring structures such as pedestrian tunnels.

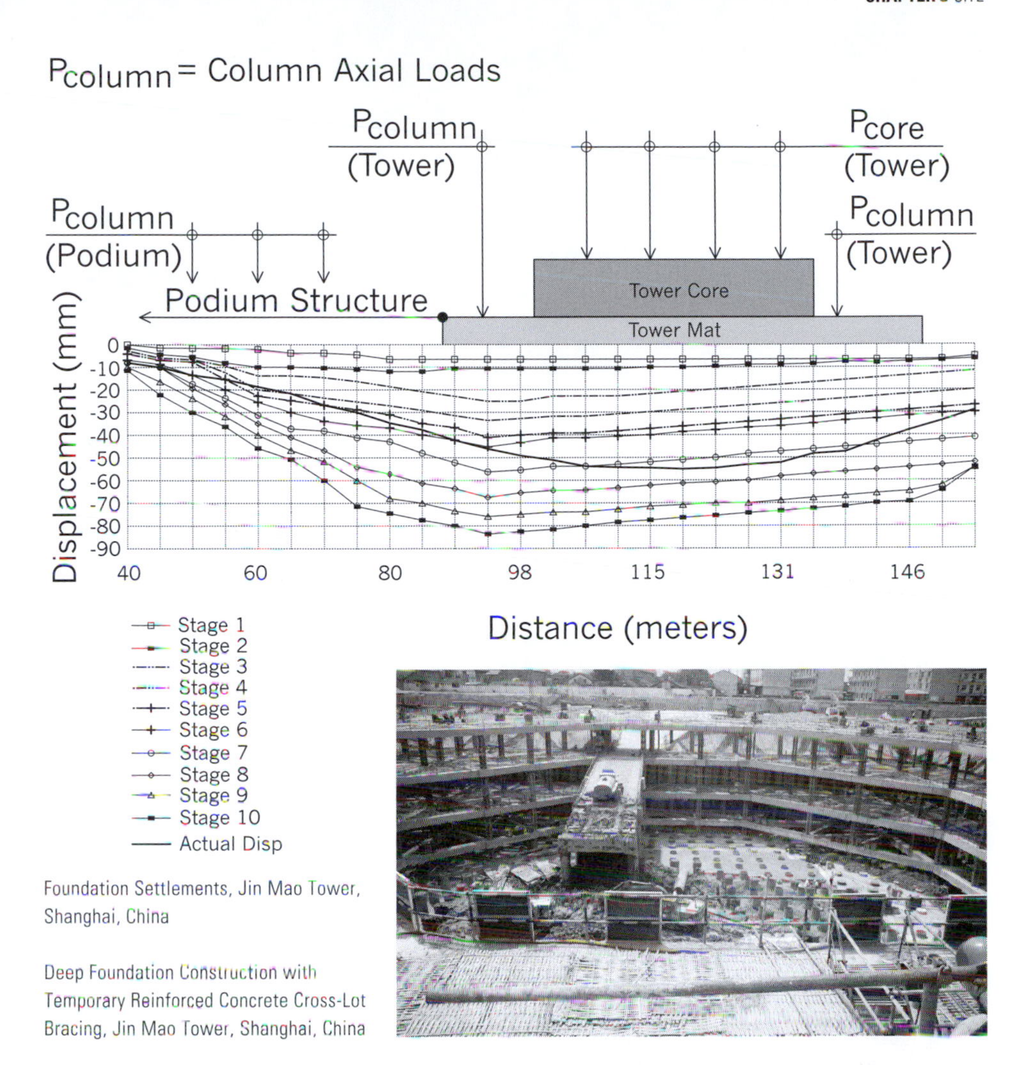

Foundation Settlements, Jin Mao Tower, Shanghai, China

Deep Foundation Construction with Temporary Reinforced Concrete Cross-Lot Bracing, Jin Mao Tower, Shanghai, China

Differential settlements of foundations are far more serious. Elastic shortening of steel/pre-cast piles and creep, shrinkage, and elastic shortening of cast-in-place piles or caissons must be considered. Uneven lengths of pile or caisson foundations require consideration for applied stress and the length subjected to sustained loads. Longer piles may need to have their cross-section oversized to control this behavior. Special site conditions during construction also must be considered. Pressure grouting of soil for stabilization or for control of ground water infiltration could result in uneven subgrade moduli. Until load is evenly distributed or forces in piles are mobilized through fracturing areas of grouting, towers may settle unevenly which could result in a serious out-of-plumb condition causing global overturning bending moments due to eccentrically placed gravity loads.

CHAPTER 4

FORCES

4.1 CODE-DEFINED LOADS

BUILDING FORCES ARE CALCULATED using loading prescribed by the applicable building code. In the United States, local building codes are based on the 2012 International Building Code (IBC), which will be presented in this chapter with other codes referenced for specific calculations. Building codes of other regions vary in format and nomenclature, but the overall theories governing the determination of forces are very similar.

4.1.1 Gravity Loads

Beyond self-weight of the structure (based on density of material used) several superimposed types of dead and live loads must be considered in design. The superimposed dead loads are attributed to partitions, ceilings, mechanical systems, floor finishes, etc., while superimposed live loads are attributed to occupancies which may vary from residential to office to retail, etc.

4.1.2 Lateral Loads

Code procedures for wind and seismic loading simply convert complex phenomena into more easily understandable calculations applicable to most buildings. For unique structures, more detailed studies such as wind tunnel testing and seismic response history analyses are used to provide more accurate results. Nevertheless, it is important to understand basic code equations before further analysis is performed.

For tall buildings, it is critical to consider both wind and seismic loading. For example, even in a high seismic region, wind load often governs the lateral design of a tall building due to the necessity of limiting building motion during high winds.

4.1.3 Risk Category

To account for the unpredictability of forces in nature, the building code assigns a risk category to each building. Buildings deemed more important for human safety and security associated with occupancy are given higher risk categories. Risk categories per the 2012 IBC are summarized as follows:

Risk Category I	Buildings that represent a low risk to human life in the event of a failure (small, seldom occupied buildings such as agricultural structures and small storage facilities)
Risk Category II	Buildings except those listed in Risk Categories I, III, and IV (most typical buildings include single-family homes and low-rise commercial buildings)
Risk Category III	Buildings that represent a substantial hazard to human life in the event of a failure (buildings with large occupancies including assembly halls, schools, jails and large residential and commercial buildings)
Risk Category IV	Buildings designated as essential facilities (buildings critical for emergency response including hospitals, fire stations and police stations; buildings containing highly toxic material)

Corresponding to each risk category are importance factors which amplify forces to provide the more critical buildings with a higher level of safety. Traditionally, snow, ice, seismic, and wind loads are given importance factors to amplify forces. However, in the most current IBC, wind loads are instead calculated based on wind speeds which vary by risk category and the wind importance factor is set to 1.0. The seismic importance factor (I_e) for each risk category per the 2012 IBC is listed below:

Risk Category I: $I_e = 1.00$

Risk Category II: $I_e = 1.00$

Risk Category III: $I_e = 1.25$

Risk Category IV: $I_e = 1.50$

4.2 CODE-DEFINED VERTICAL FORCE DISTRIBUTION FOR WIND

Wind loads as defined by the basic equation in Chapter 3.1 are the basis for wind load magnitude. Wind forces generally affect the windward (in direct path of wind) and leeward (opposite face) of the structure. Wind loads vary with height, increasing with distance from the ground. These loads must be applied to the face area of the structure and consider both windward and leeward

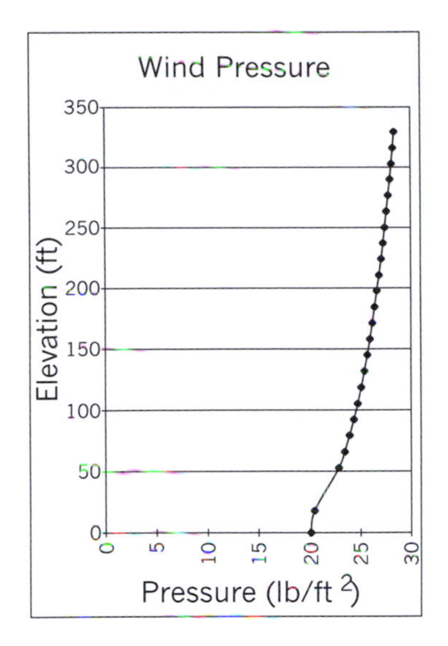

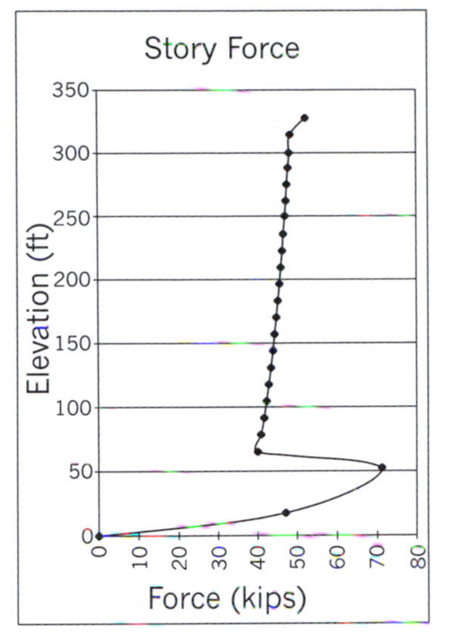

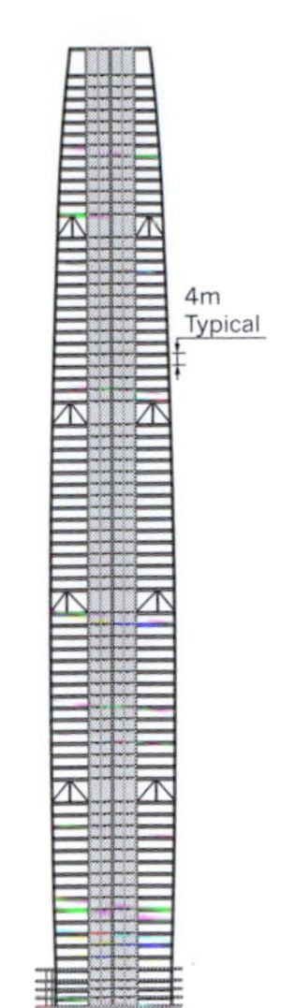

Applied Wind Pressures and Resulting Tower Story Forces

effects. The image considers a wind force distribution along the height of the tower given the site design criteria. Also included is the force distribution (based on tributary width) and resulting story forces applied to the tower.

4.3 CODE-DEFINED VERTICAL FORCE DISTRIBUTION FOR SEISMIC

The 2012 International Building Code (IBC) is the most current commonly used building code and incorporates the procedures of the 2010 American Society of Civil Engineers Minimum Design Loads for Buildings and Other Structures (ASCE 7-10). The 1997 Uniform Building Code (UBC) is referenced because it was instrumental in defining specific seismic requirements based on the latest research and the most advanced technologies at that time.

Seismic loads as defined by the IBC 2012 and the UBC 1997 are calculated as follows given the site conditions defined.

4.3.1 Earthquake Force

4.3.1.1 Earthquake Force (*E*)—IBC 2012 and ASCE 7-10

The following is the general definition of the earthquake force (*E*) that must be considered to act on a structure:

$$E = E_h \pm E_v$$

where,

E_h = earthquake force due to the base shear $(V) = \rho Q_E$

E_v = the load effect resulting from the vertical component of the earthquake motion $0.2S_{DS}D$

S_{DS} = design spectral response acceleration at short periods; calculated in section 4.3.2

D = effect of dead load

ρ = redundancy factor per ASCE 7-10 Chapter 12
= 1.0 for redundant, regular buildings complying with ASCE 7-10 checks
= 1.3 for all other buildings

Q_E = effects of horizontal forces from V, where required these forces act simultaneously in two directions at right angles to each other

When design requires the inclusion of an overstrength factor, E shall be defined as follows:

$$E_m = E_{mh} \pm E_v$$

where,

E_m = seismic load effect including overstrength factor

E_{mh} = effect of horizontal seismic forces including structural overstrength $= \Omega_o Q_E$

where,

Ω_o = seismic force amplification factor (overstrength factor)

4.3.1.2 Earthquake Force (E)—UBC 1997

The following is the general definition of the earthquake force (E) that must be considered to act on a structure:

$$E_m = \rho E_h + E_v$$

where:

E_h = earthquake force due to the base shear (V)

E_v = the load effect resulting from the vertical component of the earthquake motion $= 0.5C_a ID$

C_a = seismic coefficient

I = seismic importance factor

ρ = reliability / redundancy factor
$= 1.0 \le \rho = 2 - \dfrac{20}{\left(r_{\max}\sqrt{A_B}\right)} \le 1.50$
($\rho \le 1.25$ for special moment resisting frames)

r_{max} = maximum element–story shear ratio. For initial calculations this is the ratio of shear in each primary load resisting element. For more refined analyses this is the ratio of the design story shear in the most heavily loaded single element divided by the total design story shear

A_B = area at base of building in sq. ft.

For critical structural system elements expected to remain essentially elastic during the design ground motion to ensure system integrity:

$$E_m = \Omega_o E_h$$

where:

Ω_o = seismic force amplification factor (overstrength factor)

4.3.2 Static Force Procedure

4.3.2.1 Seismic Base Shear (*V*)—IBC 2012 and ASCE 7-10

The following is a static force procedure based on an approximate method for determining seismic base shear considering the design basis ground motion:

$$V = C_S W$$

where:

V = seismic base shear

W = the effective seismic weight

C_S = the seismic response coefficient = $\dfrac{S_{DS}}{\left(\dfrac{R}{I_e}\right)}$

where:

S_{DS} = the design spectral response acceleration parameter in the short period range =

$$\frac{2}{3} S_{MS}$$

where:

S_{MS} = the maximum considered earthquake spectral response accelerations for short period = $F_a S_s$

where:

F_a = short period site coefficient per IBC 2012 Table 1613.3.3(1)

S_s = the mapped maximum considered spectral accelerations for short periods; can be determined using IBC 2012 Figures 1613.3.1(1)–(6) or using the seismic design maps of www.usgs.gov

R = the response modification factor per ASCE 7-10 Table 12.2-1 (structural system dependent)

I_e = seismic importance factor per ASCE 7-10 Table 1.5-2

However, the value of C_s need not exceed the following:

$$C_s = \frac{S_{D1}}{T\left(\frac{R}{I_e}\right)} \text{ for } T \leq T_L$$

$$C_s = \frac{S_{D1}T_L}{T^2\left(\frac{R}{I_e}\right)} \text{ for } T > T_L$$

C_s shall not be less than: $C_s = 0.044\ S_{DS}\ I \geq 0.01$

In addition, for structures located where S_1 is equal to or greater than 0.6g, C_s shall not be less than:

$$C_s = \frac{0.5S_1}{\left(\frac{R}{I_e}\right)}$$

where:

S_{D1} = the design spectral response acceleration parameter at a period of 1.0s = $\frac{2}{3}S_{M1}$

where:

S_{M1} = the maximum considered earthquake spectral response accelerations for 1-second period = $F_V S_1$

where:

F_V = long period site coefficient

S_1 = the mapped maximum considered spectral accelerations for a 1-second period; can be determined using IBC 2012 Figures 1613.3.1(1)–(6) or using the seismic design maps of www.usgs.gov

T = the fundamental period of the structure (s)

T_L = long-period transition period (s) per ASCE7-10 Figures 22-12–22-16

4.3.2.2 Fundamental Period (Approximate Methods)—IBC 2012 and ASCE 7-10

The fundamental period, T, of a structure is dependent on the building height and stiffness and can be determined analytically. However, it is also acceptable to estimate the period using the following equation:

$$T_a = C_t(h_n)^x$$

where:

T_a = approximate fundamental period
C_t = 0.028, $x = 0.8$ for steel moment-resisting frames
C_t = 0.016, $x = 0.9$ for concrete moment-resisting frames
C_t = 0.03, $x = 0.75$ for steel eccentrically braced frames and buckling-restrained braced frames
C_t = 0.02, $x = 0.75$ for all other structural systems
h_n = height from the base of the building to the highest level (feet)

When the period of a structure is calculated by a permitted analysis model, the period used for determination of base shear must be limited to $T_a \times C_u$, where C_u is the coefficient for upper limit on calculated period per ASCE 7-10 Table 12.8-1.

Alternatively, it is permitted to determine the approximate fundamental period (T_a) in seconds from the following equation for structures not exceeding 12 stories in height in which the seismic force-resisting system consists entirely of concrete or steel moment-resisting frames and the story height is at least 3m (10ft):

$$T_a = 0.1N$$

where:

N = number of stories

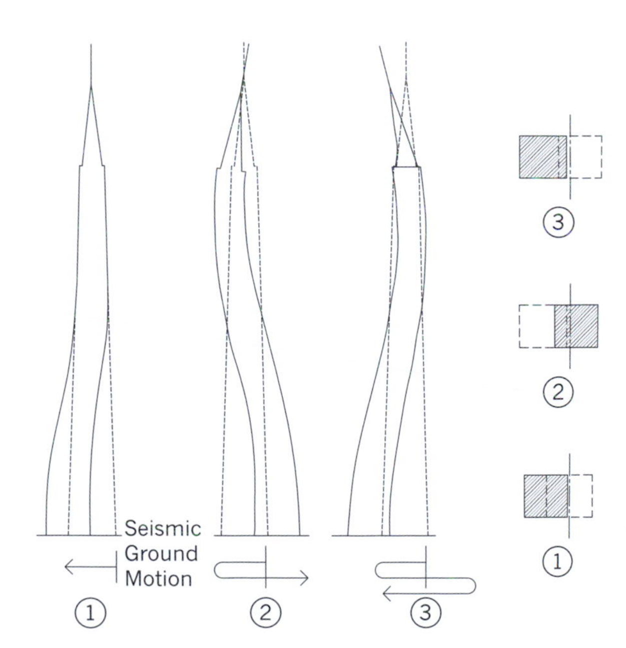

Tower Motion Resulting from Seismic Ground Motion, Elevation and Plan Diagrams

The approximate fundamental period, T_a, for masonry or concrete shear wall structures is permitted to be determined as follows:

$$T_a = \frac{0.0019h_n}{\sqrt{C_W}}$$

where:

$$C_W = \frac{100}{A_B}\sum_{i=1}^{x}\left(\frac{h_n}{h_i}\right)^2 \frac{A_i}{\left[1+0.83\left(\frac{h_i}{D_i}\right)^2\right]}$$

where:

A_B = area of base of structure (sq ft)
A_i = web area of shear wall "i" (sq ft)
D_i = length of shear wall "i" (feet)
h_i = height of shear wall "i" (feet)
X = number of shear walls in the building effective in resisting lateral forces in the direction under consideration

4.3.2.3 Seismic Dead Load (*W*)—IBC 2012 and ASCE 7-10

Applicable portions of other loads to be considered for the total seismic load, *W*, include:

1. In storage and warehouse occupancies, a minimum of 25% of the floor live load.
2. Where partition loads are used in floor design, a load not less than 10 psf.
3. Total operating weight of permanent equipment.
4. Where the flat roof snow load exceeds 30 psf, 20% of the uniform design snow load regardless of actual roof slope.

4.3.2.4 Seismic Base Shear (*V*)—UBC 1997

The following is a static force procedure based on an approximate method for determining base shear considering the design basis ground motion:

$$\frac{2.5C_aI}{R}W \geq V = \frac{C_vI}{RT}W \geq 0.11C_aIW$$

$$\geq \frac{0.8ZN_vI}{R}W \text{ (Seismic Zone 4)}$$

where:

C_a and C_v = seismic coefficients
I = seismic importance factor
W = total dead load plus applicable portions of other loads

R = response modification factor
T = fundamental period of vibration of the structure
Z = seismic zone factor
N_v = velocity-dependent near-source factor

4.3.2.5 Seismic Dead Load (*W*)—UBC 1997

Applicable portions of other loads to be considered for the total seismic load, *W*, include:

1. In storage and warehouse occupancies, a minimum of 25% of the floor live load.
2. Where partition loads are used in floor design, a load not less than 10 psf.
3. 20% of the uniform design snow load, when it exceeds 30 psf.
4. Total weight of permanent equipment.

4.3.2.6 Fundamental Period (Approximate Methods)—UBC 1997

For the determination of building period (T), by the approximate Method A:

$$T = C_t(h_n)^{3/4}$$

C_i = 0.035 for steel moment-resisting frames
C_i = 0.030 for reinforced concrete moment-resisting frames and eccentrically braced frames
C_i = 0.020 for other concrete buildings

where:

h_n = height from the base of the building to the highest level (feet)

Alternatively, for structures with concrete or masonry shear walls:

$$C_t = \frac{0.1}{\sqrt{A_c}}$$

where:

A_c = combined effective area of shear walls in the first story of the structure (sq ft) =

$$\sum A_e\left[0.2 + (D_e / h_n)^2\right],\ \frac{D_e}{h_n} \leq 0.9$$

A_e = minimum cross sectional area in any horizontal plane in the first story of a shear wall (sq ft)
D_e = length of shear wall in the first story in the direction parallel to the applied forces (feet)

Once preliminary sizes are obtained based on the base shear calculated using the approximate period *T*, a more accurate value of *T* can be determined using established analytical procedures.

In lieu of approximate Method A, Method B provided in the UBC Code can be used to determine *T*. Method B permits the evaluation of *T* by either the Rayleigh formula or other substantiated analysis. Note that the value of *T* obtained from Method B must be less than or equal to 1.3 times the value of *T* obtained from Method A in Seismic Zone 4, and less than or equal to 1.4 times the value in Seismic Zones 1, 2, and 3.

4.3.3 Distribution of Lateral Forces

4.3.3.1 Vertical Force Distribution

The base shear (*V*) is distributed vertically to each floor level of the building. The story shears are then distributed to the lateral force resisting elements proportional to their relative stiffness and the stiffness of the diaphragms.

As described in the UBC 1997, the base shear is distributed linearly over the height of the building, varying from zero at the bottom to a maximum at the top, corresponding to the fundamental (first mode) period of vibration of the structure. To account for higher mode effects on the structure (important for buildings with a fundamental period greater than 0.7 seconds), a portion of the base shear is applied as a concentrated load at the top of the building (see Section 4.3.3.2). Distribution of forces is similar in the IBC 2012 to the UBC procedure. However, in accounting for higher mode effects, the IBC increases story forces exponentially up the height of the building rather than applying an additional point load.

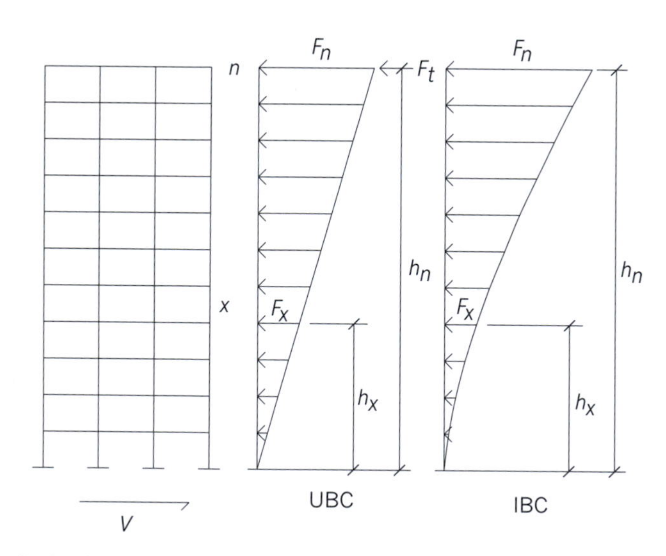

Vertical Force Distribution for Seismic Landing

4.3.3.2 Force Distribution Procedure—UBC 1997

The seismic design story shear in any story (V_x) is determined as follows:

$$V_x = F_t + \sum_{i=x}^{n} F_i$$

and the base shear is:

$$V = F_t + \sum_{i=1}^{n} F_i$$

where:

F_i, F_n, F_x = design seismic force applied to level i, n, or x, respectively
F_t = portion of V considered concentrated at the top of the structure in addition to F_n
$F_t = 0.07TV < 0.25V$ for $T > 0.7$ seconds
= 0.0 for $T > 0.7$ seconds

and,

$$F_x = \frac{(V - F_t) w_x h_x}{\sum_{i=1}^{n} w_i h_i}$$

where:

h_i h_x = height above the base to level i or x, respectively
w_i w_x = the portion of W located at or assigned to level i or x, respectively
T = fundamental period of vibration of structure in seconds in direction of analysis

4.3.3.3 Force Distribution Procedure—IBC 2012 and ASCE 7-10

The seismic design story shear in any story (V_x) is determined as follows:

$$V_x = \sum_{i=x}^{n} F_i$$

and the base shear is:

$$V = \sum_{i=1}^{n} F_i$$

where:

$$F_x = \frac{V w_x h_x^k}{\sum_{i=1}^{n} w_i h_x^k}$$

where:

$F_i F_n F_x$ = design seismic force applied to level *i, n,* or *x,* respectively

$h_i h_x$ = height above the base to level *i* or *x,* respectively

$w_i w_x$ = the portion of *W* located at or assigned to level *i* or *x,* respectively

k = 1 for $T \leq 0.5$s

k = 2 for $T \geq 2.5$s

k is determined by linear interpolation between 1 and 2 for $0.5 < T < 2.5$s

4.3.4 Bending Moment Distribution (Overturning)

Once design seismic forces applied to levels have been established, the bending moment due to these forces can be determined. The tower structure must be designed to resist the overturning effects caused by the earthquake forces. The overturning moment (M_x) at any level x can be determined by the following formula:

$$M_x = \sum_{i=x}^{n} F_i(h_i - h_x) + F_t(h_n - h_x)$$

where:

F_i = portion of seismic base shear (V) located or assigned to level i

F_i = portion of V considered concentrated at the top of the structure in addition to F_n (= 0 for IBC 2012 procedure)

$h_i h_n h_x$ = height in ft above the base to level *i, n,* or *x,* respectively

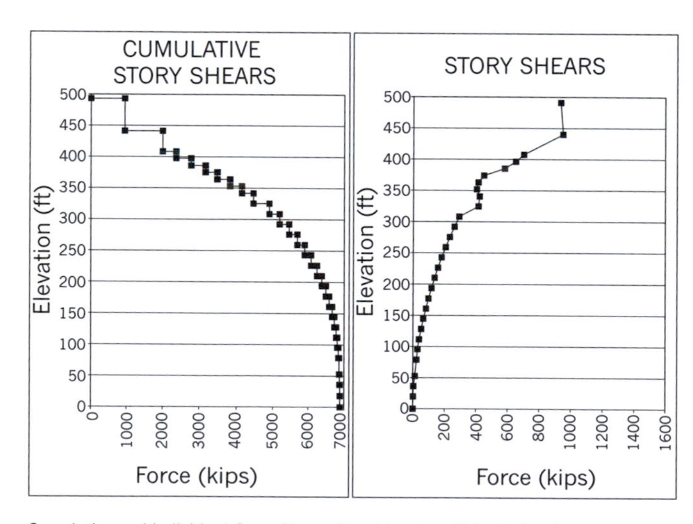

Cumulative and Individual Story Shears Resulting from Seismic Loading

4.3.5 Story Drift Limitations

Drift is defined as the displacement that a structure experiences when subjected to load. Drift is usually caused by lateral loads due to wind or seismic events but can be caused by unbalanced gravity loads or temperature effects disproportionately applied. The structure experiences overall drift which is described as the displacement at the top of the building relative to the ground.

Inter-story drift is the relative displacement of one floor level to another. For seismic events this calculation is important because inter-story drifts due to inelastic response could be large. Exterior wall and partition systems among other vertical building systems must be detailed to allow for this movement.

Inter-story drifts within the structure shall be limited to a maximum inelastic drift response, approximately equal to the displacement that occurs in the structure when subjected to the design basis ground motion. In UBC 1997, expected drift is calculated using the following equation:

$$\Delta_M = 0.7R\Delta_S$$

where:

Δ_M = maximum expected inelastic drift
R = response modification factor
Δ_S = maximum computed elastic drift considering the lateral force resisting system

For structures with a fundamental period (T) less than 0.7 seconds, the calculated story drift using Δ_M shall not exceed 0.025 (2.5%) the story height. For structures with (T) greater than or equal to 0.7 seconds, the story drift shall not exceed 0.02 (2%) times the story height.

IBC 2012 uses a slightly modified equation to predict inelastic drift:

$$\Delta_e = (C_d/I_e)\Delta_{xe}$$

where:

Δ_e = maximum expected inelastic drift
C_d = deflection amplification factor per ASCE 7-10 Table 12.2-1 (structural system dependent)
Δ_{xe} = maximum computed elastic drift
I_e = seismic importance factor

Allowable story drift (Δ_a) per IBC 2012 is defined in ASCE 7-10 Table 12.12-1 and is dependent on building system, occupancy and height. For most buildings above 4 stories, Δ_a = 0.020 (2%) times the story height for Risk Categories I and II, 0.015 (1.5%) times the story height for Risk Category III and 0.010 (1%) times the story height for Risk Category IV. For shorter buildings, the limits are more lenient.

4.4 GRAVITY LOAD DISTRIBUTION AND TAKEDOWNS

4.4.1 Floor Systems

Gravity loads are generally considered to be uniformly distributed over an occupied floor. These loads vary based on building use and include dead load (self-weight), superimposed dead load (load from building components that have little variation in magnitude of load over the life of the building, i.e. partitions, ceiling systems, mechanical systems), and live load (that can vary in magnitude and location).

When considering dead load (self-weight), all components of the primary structure must be included, typically consisting of floor slabs, floor framing beams and girders, and columns. Material density must be known in order to accurately calculate self-weight of the structure. Common densities include structural steel 7850 kg/cu m (490 lbs/cu ft) and reinforced concrete 2400 kg/cu m (150 lbs/cu ft).

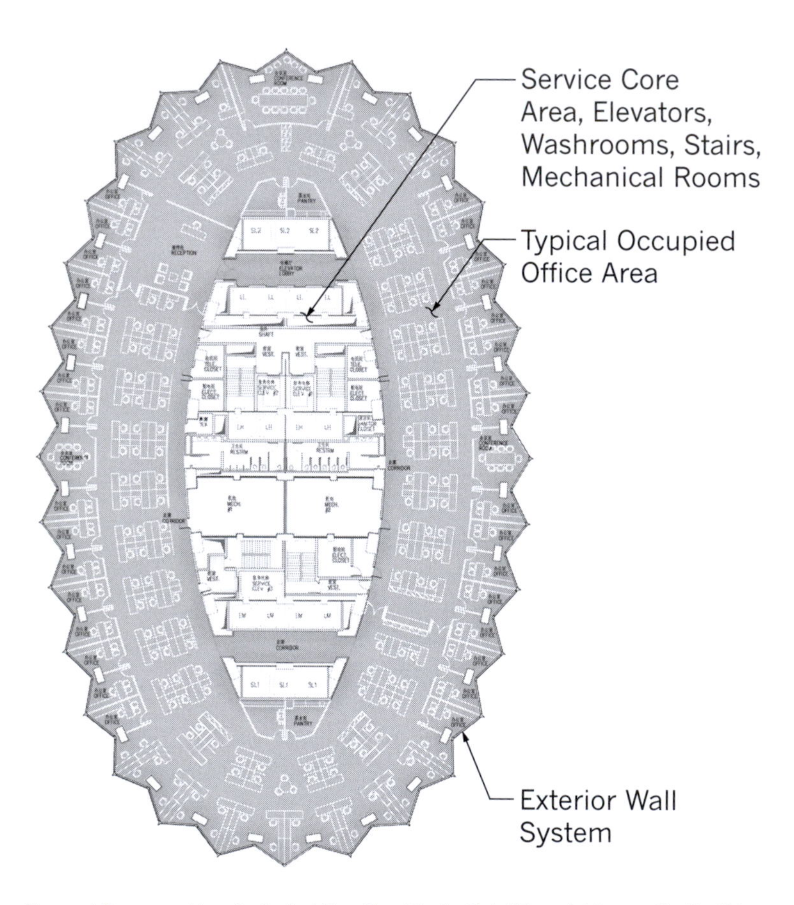

Proposed Occupancy Use of a Typical Floor Plan, Tianjin Global Financial Center, Tianjin, China

Common superimposed dead and live loads as defined by codes for a building structure are listed as follows:

Superimposed Dead Load (SDL):

Partitions (dry wall)	= 1.0 kPa (20 psf)
Ceiling (panel system)	= 0.15 kPa (3 psf)
Mechanical systems	= 0.10 kPa (2 psf)
Library or storage	= 7.5 kPa (150psf)
Finished flooring	= 1.2 kPa (25 psf)

Live Loads (LL):

Office	= 2.5 kPa (50 psf)
Office (premium)	= 4.0 kPa (80 psf)
Residential/hotel	= 2.0 kPa (40 psf)
Public spaces (i.e. lobbies)	= 5.0 kPa (100 psf)
Parking (passenger veh.)	= 2.0 kPa (40 psf)

There are cases where individual concentrated point or specific line loads must be considered on the structure. An example of a concentrated point load may include the consideration of truck loadings in a loading dock, and an example of specific line load may include the consideration of a heavy masonry partition used for acoustic isolation.

4.4.2 Exterior Walls

Exterior walls for tower structures produce specific loads that must be considered in the structure design. These loads may be light (e.g., metal panels and glass) or heavy (e.g., precast concrete) depending on the

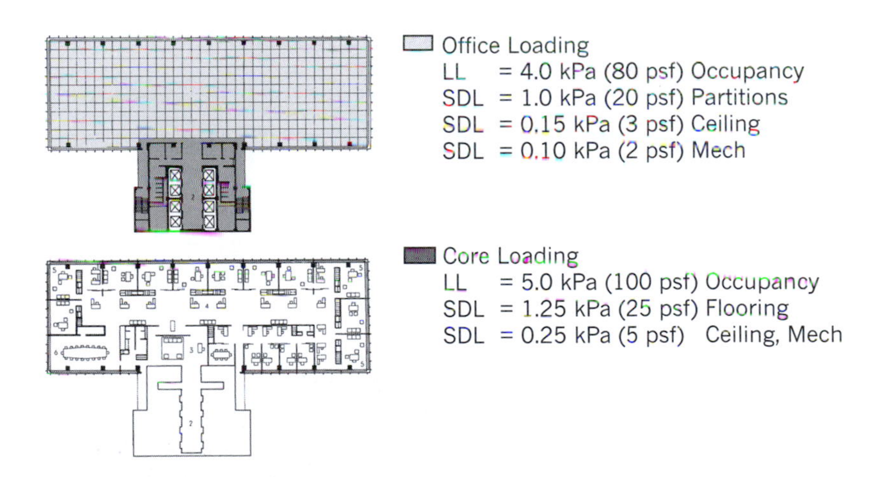

Design Floor Loading, Typical Office Floor Plan

architectural design. Based on exterior wall connections, imposed loads can be calculated. In some cases, the exterior wall is supported on perimeter girders and in some cases directly to columns. For initial calculations, exterior wall loads may be considered evenly distributed along perimeter spandrels, considering the exterior wall weight and the floor-to-floor heights.

Exterior wall loads are often considered as distributed load over the "face" area of the structure. Some common exterior wall loads are as follows (all of which would need to be confirmed based on the final as-designed exterior wall system):

Metal and Glass = 0.75 kPa (15 psf)
Stone and Glass = 1.2 kPa (25 psf)
Precast and Glass = 2.5 kPa (50 psf)

4.4.3 Loads to Vertical Elements

Loads, either distributed or concentrated, are first typically supported by horizontal framing members, then by vertical columns or walls, then by foundation systems.

Loads distributed to floor slabs are typically supported by beam framing, then girder framing, then to columns or walls. Knowing the required spans and support conditions of the floor framing elements, distributed loads are used for the design of the members. These loads then transfer through the horizontal support systems to vertical load carrying elements. Generally, columns and walls support a tributary floor area, as well as a tributary exterior wall area with the deletion of any floor or wall openings. A column or wall load takedown is performed on each discrete column or wall element from

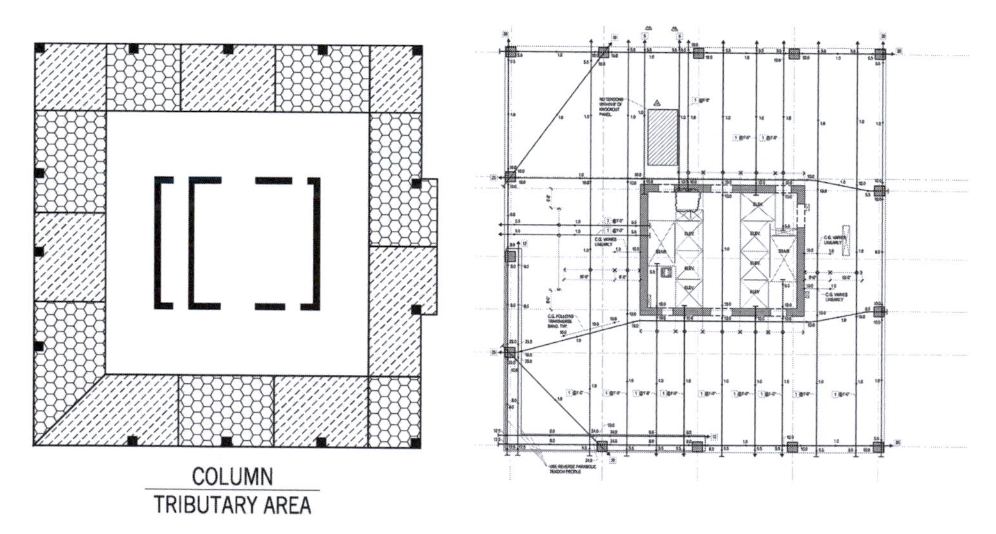

Tributary Loading Area to Columns, Floor Framing Plan, 350 Mission Street, San Francisco, CA

the influence of multiple floors on the elements. Codes typically recognize a reduction in live load on vertical elements when multiple floors and large areas are considered.

4.5 LOAD COMBINATIONS

In accordance with IBC 2012, the following symbol definitions and load combinations shall be used:

D = dead load
E = combined effect of horizontal and vertical earthquake induced forces as defined in Section 4.3 (or Section 12.4.2 of ASCE 7-10)
E_m = maximum seismic load effect of horizontal and vertical seismic forces as set forth in Section 4.3 (or Section 12.4.3 of ASCE 7-10)
F = load due to fluids with well-defined pressures and maximum heights
H = load due to lateral earth pressure, ground water pressure, or pressure of bulk materials
L = live load, except roof live load, including any permitted live load reduction
L_r = roof live load including any permitted live load reduction
R = rain load
S = snow load
T = self-straining force arising from contraction or expansion resulting from temperature change, shrinkage, moisture change, creep in component materials, movement due to differential settlement or combinations thereof (the effects of self-straining forces should always be considered, but their application in the following combinations will be determined on a case-by-case basis using engineering judgment; the factor applied to T will often match that applied to D)
W = load due to wind pressure

4.5.1 Basic Load Combinations—Strength or Load and Resistance Factor Design

1. $1.4(D + F)$
2. $1.2(D + F) + 1.6(L + H) + 0.5(L_r \text{ or } S \text{ or } R)$
3. $1.2(D + F) + 1.6(L_r \text{ or } S \text{ or } R) + 1.6H + (f_1 L \text{ or } 0.5W)$
4. $1.2(D + F) + 1.0W + f_1 L + 0.5(L_r \text{ or } S \text{ or } R)$
5. $1.2(D + F) + 1.0E + f_1 L + f_2 S$ or
   $(1.2 + 0.2S_{DS})D + \rho Q_E + L + 0.2S$ or

$(1.2 + 0.2S_{DS})D + \Omega_o Q_E + L + 0.2S$—when considering structural overstrength

6. $0.9D + 1.0W + 1.6H$
7. $0.9(D + F) + 1.0E + 1.6H$ or $(0.9 - 0.2S_{DS})D + \rho Q_E + 1.6H$
   $(0.9 - 0.2S_{DS})D + \Omega_o Q_E + 1.6H$—when considering structural overstrength

Where:

f_1 = 1 for floors in places of public assembly, for live loads in excess of 100psf , and for parking garage live load, and = for other live loads

f_2 = 0.7 for roof configurations (such as saw tooth) that do not shed snow off the structure, and = 0.2 for other roof configurations

Exception: Where other factored load combinations are specifically required by the provisions of the IBC 2012 Code, such combinations shall take precedence.

4.5.2 Basic Load Combinations—Allowable (Working) Stress Design

1. $D + F$
2. $D + H + F + L$
3. $D + H + F + (L_r \text{ or } S \text{ or } R)$
4. $D + H + F + 0.75(L) + 0.75(L_r \text{ or } S \text{ or } R)$
5. $D + H + F + (0.6W \text{ or } 0.7E)$ or
   $(1.0 + 0.14S_{DS})D + H + F + 0.7\rho Q_E$ or
   $(1.0 + 0.14S_{DS})D + H + F + 0.7\Omega_o Q_E$
   when considering structural overstrength
6. $D + H + F + 0.75(0.6W \text{ or } 0.7E) + 0.75L + 0.75(L_r \text{ or } S \text{ or } R)$ or
   $(1.0 + 0.105S_{DS})D + H + F + 0.525\rho Q_E + 0.75L + 0.75(L_r \text{ or } S \text{ or } R)$
   $(1.0 + 0.105S_{DS})D + H + F + 0.525\Omega_o Q_E\ 0.75L + 0.75(L_r \text{ or } S \text{ or } R)$
   when considering structural overstrength
7. $0.6D + 0.6W + H$
8. $0.6(D + F) + 0.7E + H$ or $(0.6 - 0.14S_{DS})D + 0.7\rho Q_E + H$
   $(0.6 - 0.14S_{DS})D + 0.7\Omega_o Q_E + H$
   when considering structural overstrength

4.6 DESIGN AXIAL, SHEAR, AND BENDING MOMENTS

Once the gravity and lateral loads along with the controlling load combinations have been established, the design axial, shear, and bending moments both globally and on individual structural elements can be established. These loads will be used in the design of the structure based on material types and behavior.

For the previous figure, design axial demands, shear demands, moment demands, and elastic deflection can be calculated according to the following equations:

$$P_A = P_R$$

$$M_{WA} = (W_A \times B)(H)\left(\frac{H}{2}\right)$$

$$M_{WA} = M_{WR}$$

$$V_{WR} = P_{WA} \times H$$

$$\delta = \frac{(W_A \times B)(H)^4}{8E_S I_S}$$

where:

P_a = applied axial gravity loading due to D, L, SDL
P_r = axial reactions due to gravity loading
M_{WA} = bending moment due to applied lateral loading (usually wind or seismic)
M_{WR} = bending moment reaction due to applied lateral loading
V_{WA} = shear due to applied lateral loading
V_{WR} = shear reaction due to applied lateral loading
E_S = modulus of elasticity of structural system
I_S = moment of inertia of structural system
δ = elastic deflection at the top of the structure

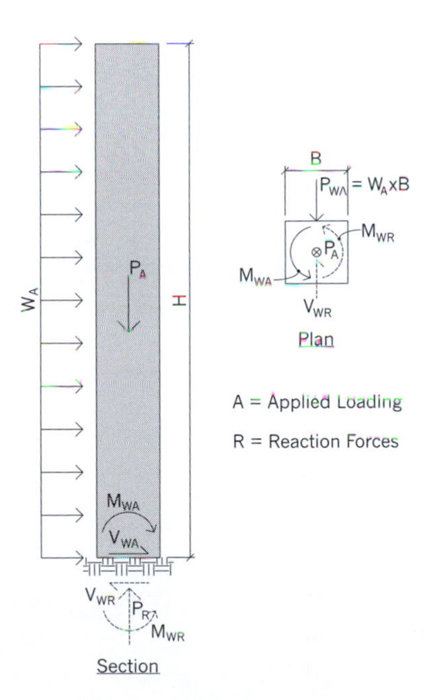

Applied Load and Resulting Forces/Reactions on Tower

CHAPTER 5

LANGUAGE

5.1 FORCE FLOW

ONE OF THE MOST IMPORTANT, and likely the most obscure considerations, is reading the force flow through a structure. An accurate understanding of this flow leads to the most correct assessment of behavior, the safest, and the most efficient design. These forces primarily originate from gravity loads and lateral loads caused by wind and seismic events. Other forces may be caused by settlement, temperature, or relative displacements of vertical columns or walls due to creep, shrinkage, and elastic shortening, etc. Once these loads acting on the structure are fundamentally understood, the flow of these forces must be understood. Forces typically flow through floor framing systems into vertical elements such as columns or walls into foundations.

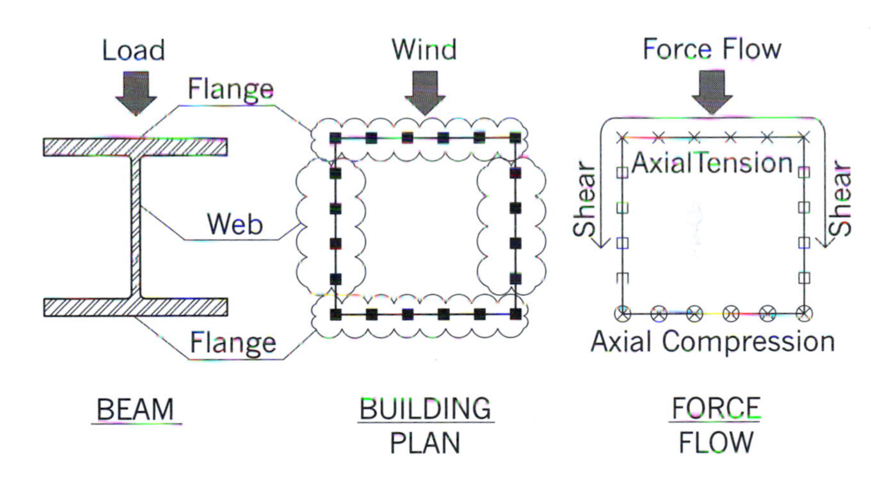

Wide-Flanged Beam Analogy to Tower Plan

FACING PAGE
John Hancock Center,
Chicago, IL

Force Flow, Tower Under Construction, John Hancock Center, Chicago, IL

Lateral force-resisting systems are typically subjected to temporary loads, externally imposed, with flow through vertical systems that are supported by foundation systems. Lateral force-resisting systems often resist gravity loads in addition to temporarily imposed lateral loads. These gravity loads, when strategically placed, actually can work to an advantage in the structure, acting as a counterbalance to overturning effects and applying "prestress" to members that would otherwise be subjected to tension when subjected to lateral loads.

Objects in nature are designed so that the least energy is expended when work is done by them. For structures to be most efficient, minimum strain energy will result in the minimum expenditure of energy to resist loads. In minimizing the energy, forces and deformations should be distributed as evenly as possible throughout the structure through a synergetic placement of material. Forces will flow through the easiest and shortest load path natural to the structure's form. Dr. Fazlur Khan sought to define natural load path to create systems demonstrating optimal performance.

Khan was empathetic to structural behavior and the dichotomy between emotional mysticism and scientific rationalism. Khan revealed that he often felt that he was himself the building when he was designing a building project.

Fazlur Khan and Bruce Graham developed three elegant designs that demonstrate proportioned structure with the flow of force. The exterior reinforced concrete columns in the 52-story One Shell Plaza, Houston, Texas (1971) were proportioned to resist gravity loads and control relative creep deformations between the core and exterior columns. These columns were enlarged toward the outside of the building rather than the inside with spandrel beams in the perimeter tubular frame designed with variable widths in plan to match the column depths. Spandrel beam depths, also proportional, decreased over the height.

Instead of introducing a depth transfer girder system for some tower columns above the lobby space at the base of the 26-story Two Shell Plaza, Houston, Texas (1972), Khan and Graham created a natural load path of reinforced concrete arches where deep spandrel beams carry vertical loads through shear with face dimensions of columns remaining constant over the height of the building.

The final example of force flow is illustrated in the 21-story Marine Midland Bank where reinforced concrete column sizes gradually increase as load is distributed to just a few columns at the lobby level. Perimeter tubular spandrel beams increase in depth in an upward manner in elevation. The columns and beams near areas of load transfer increase in dimension both in

One Shell Plaza, Houston, TX

Two Shell Plaza, Houston, TX

Marine Midland Bank,
Rochester, NY

and out of plan. Column face dimensions are almost impossibly thin over the height of the building—particularly at the top of the building at the mechanical penthouse. Columns were removed from the corner where Graham conceived of the architecture as a demonstration of how a building should land on its base.

5.2 STRUCTURAL FRAMING PLANS

5.2.1 Lateral vs. Gravity Systems

Lateral and gravity system components often are not defined specifically in a framing plan. These framing plans typically only show a portion of the structure with vertical component generally indicated but more completely described in overall structural system elevations and sections.

In the definition of force flow, the evaluation usually begins in plan, with elevations and sections following, and foundation systems last. Forces superimposed on floors migrate from elements with least stiffness to those with greatest stiffness. The typical migration originates in slab framing to floor beams or trusses, to floor girders, to columns or walls. This behavior is the same for all materials, noting that in some structures floor beams or girders may not exist with systems only incorporating flat slabs or "plates."

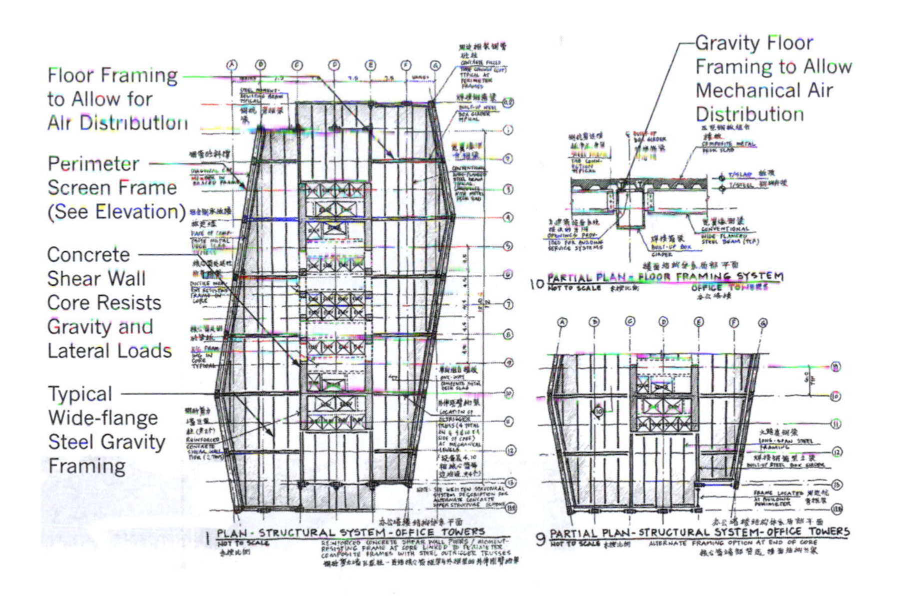

Tower Structural System Plans and Detail, Hang Lung Competition, China

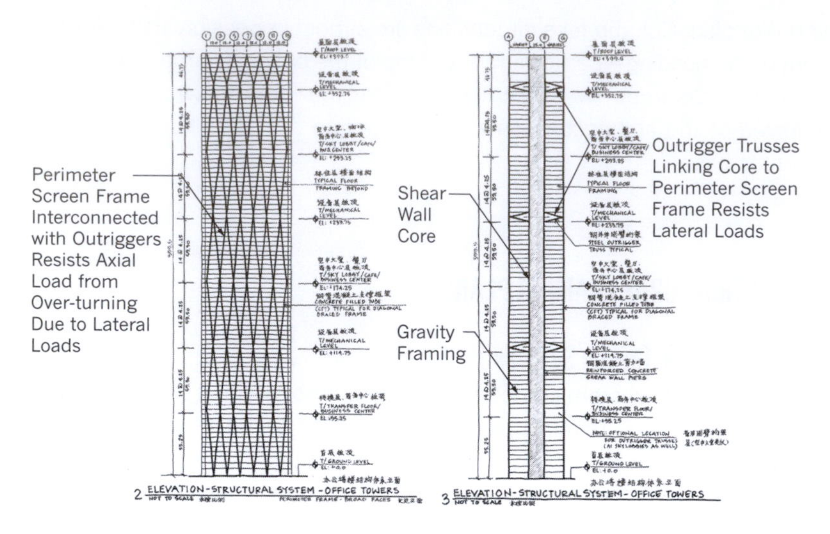

Tower Structural System Elevation and Section, Hang Lung Competition, China

5.2.2 Steel

John Hancock Center, Chicago, Illinois—Structural steel floor framing spans from the interior structural steel columns resisting gravity loads only (typical connections from beam framing to interior columns consist of shear web connections only without flange connections) to exterior braced frame where connections of diagonal, vertical, and horizontal members develop the full moment capacity of the members. The exterior braced tubular frame resists gravity and lateral loads.

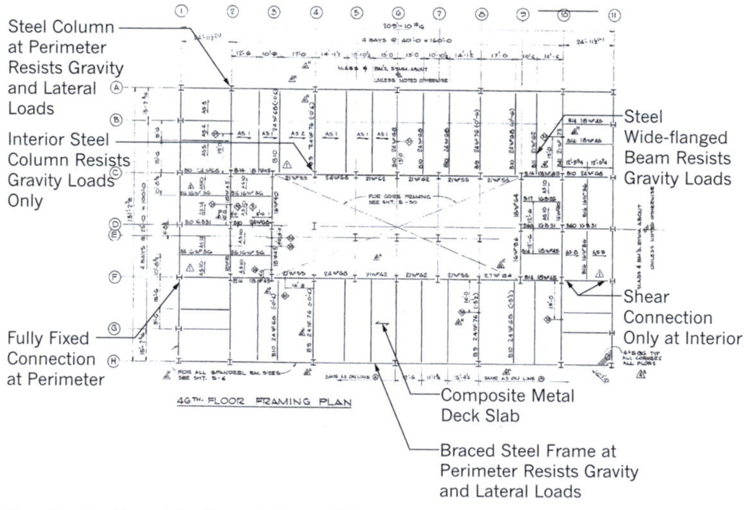

Floor Framing Plan, John Hancock Center, Chicago, IL

222 Main, Salt Lake City, Utah—Structural steel floor framing spans from the interior concentric braced frames (unbonded braces) to perimeter moment-resisting frames (beam-to-column connections are fully welded to develop full moment and shear capacity of the joints).

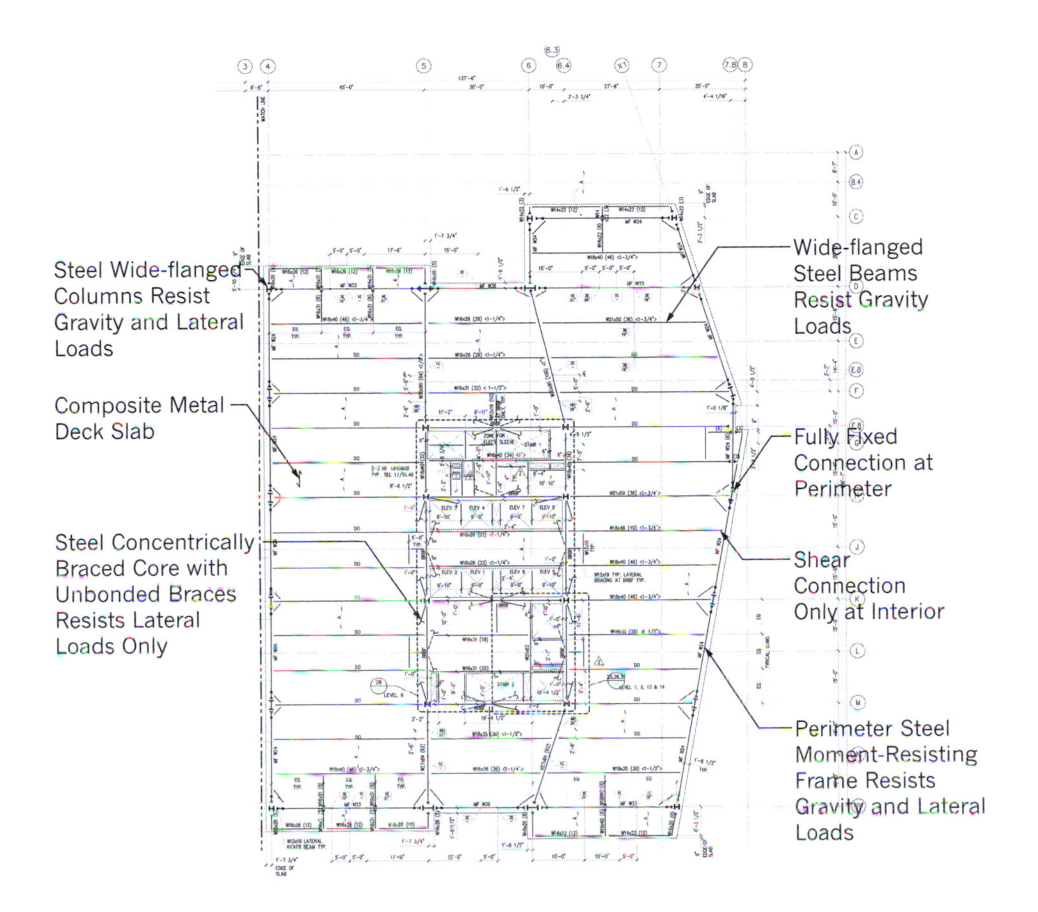

Floor Framing Plan, 222 South Main, Salt Lake City, UT

5.2.3 Concrete

500 West Monroe, Chicago, Illinois—Long-span post-tensioned concrete framing spans from the perimeter reinforced concrete frames to the central reinforced concrete shear wall core. The perimeter frames and the central core resist both lateral and gravity loads.

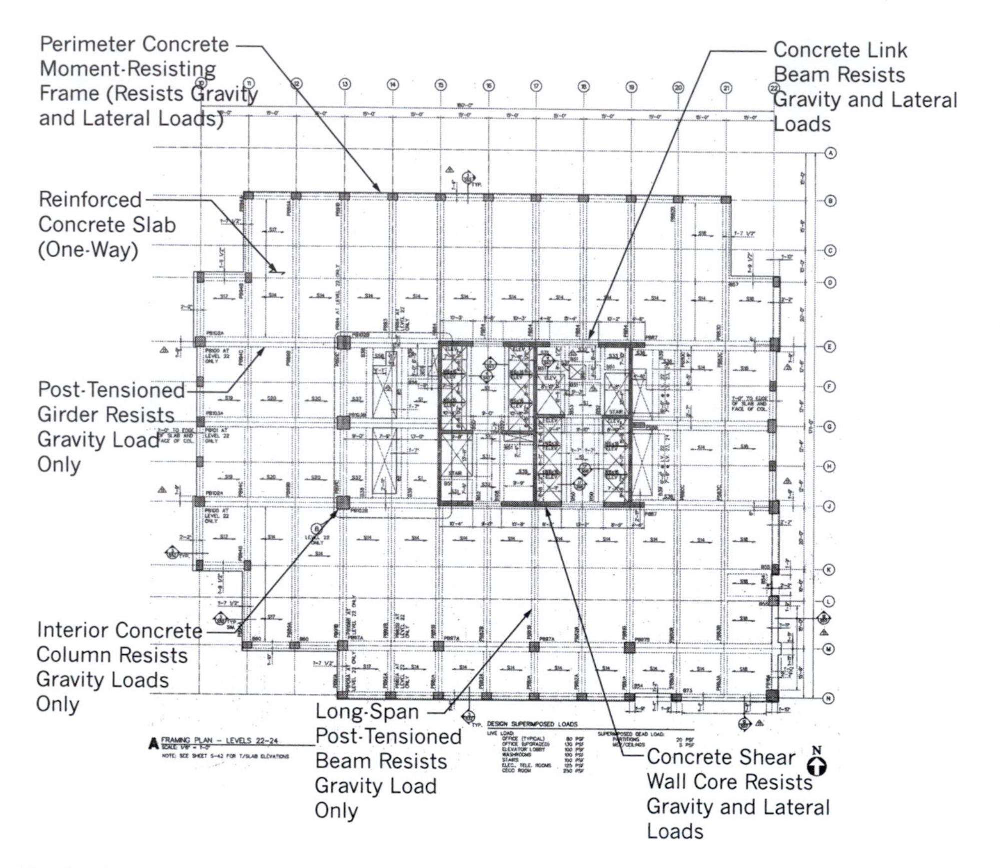

Floor Framing Plan, 500 West Monroe, Chicago, IL

University of California, Merced, Library, California—Conventional long-span reinforced concrete framing spans from internal reinforced concrete frames in each primary building direction.

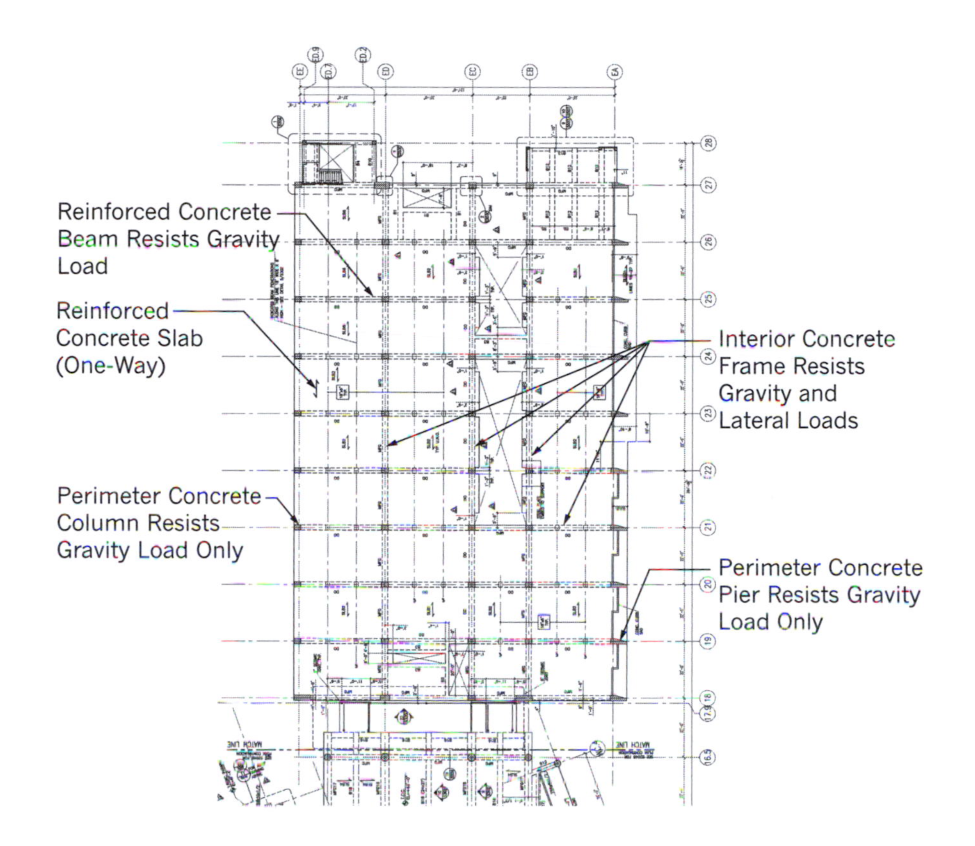

Floor Framing Plan, University of California, Merced—Kolligian Library, Merced, CA

5.2.4 Composite

China Poly Headquarters, Beijing, China—Structural steel floor framing spans from central reinforced concrete shear wall cores to perimeter steel moment-resisting frames. End connections for typical framing members use bolted shear tab connections with perimeter frame connections using a combination of bolts and welds to fully develop moment capacity of frame members. The shear wall cores and perimeter frames resist both gravity and lateral loads.

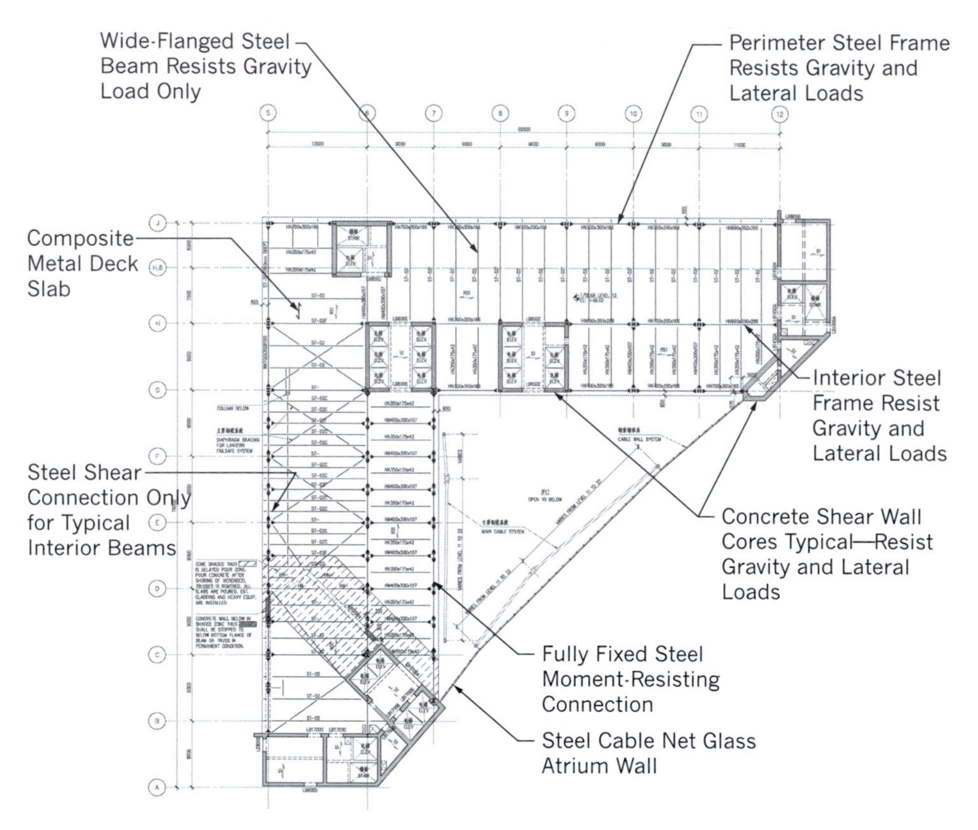

Floor Framing Plan, Poly Corporation Headquarters, Beijing, China

Kingtown International Center, Nanjing, China—Reinforced concrete long-span conventional framing spans between central reinforced concrete shear wall core and perimeter reinforced concrete tubular frame. Diagonal structural steel tubes are located on the outside of the perimeter tube to provide increased lateral load resistance. The central core, perimeter frame, and the diagonal braces all provide lateral load resistance, while the core and perimeter frame also resist gravity loads.

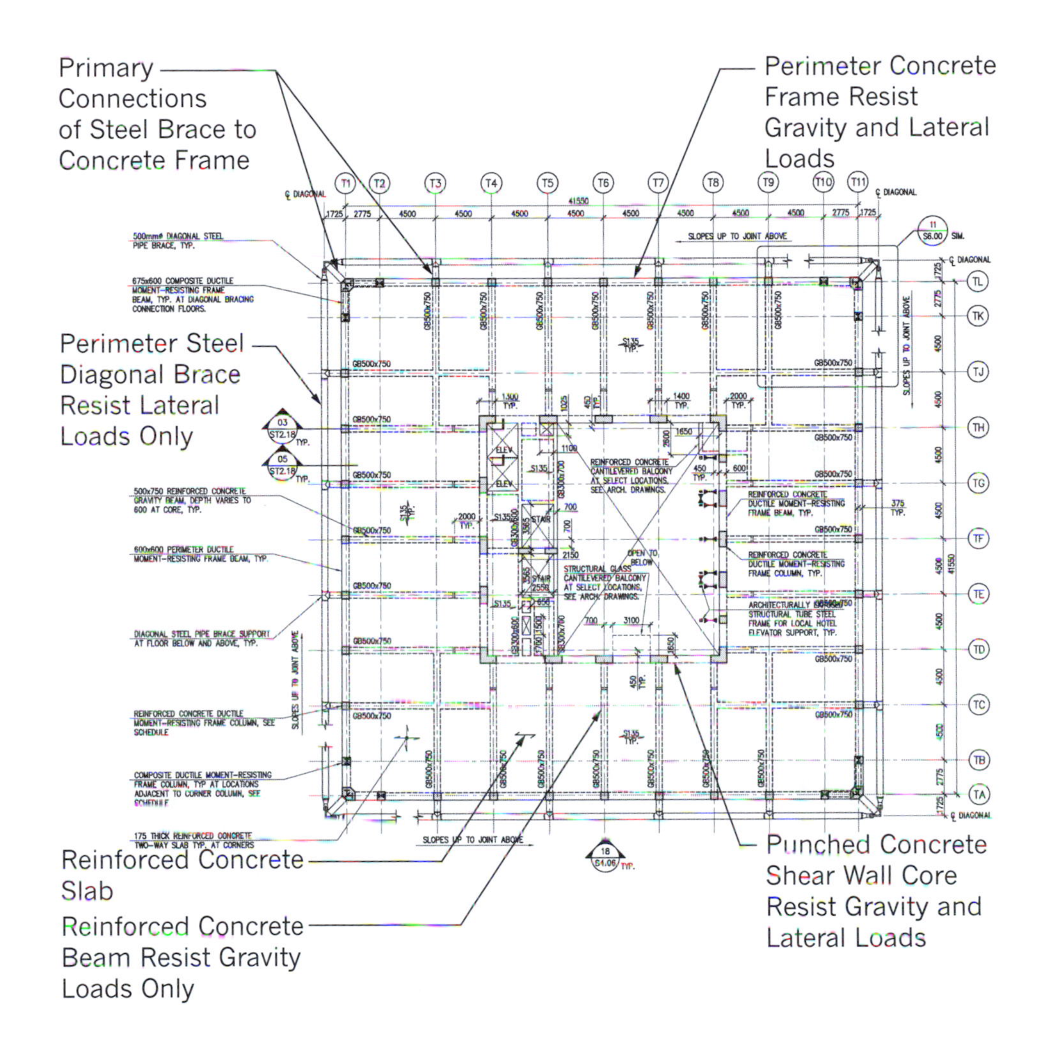

Floor Framing Plan, Kingtown International Center, Nanjing, China

5.3 STRUCTURAL SYSTEM ELEVATIONS

5.3.1 Steel

Willis Tower (formally Sears Tower), Chicago, Illinois—Bundled steel tubular frame is used to resist lateral and gravity loads. Frames exist at the perimeter and in internal locations. Steel belt trusses are used to transfer lateral loads when tubular frame steps in elevation.

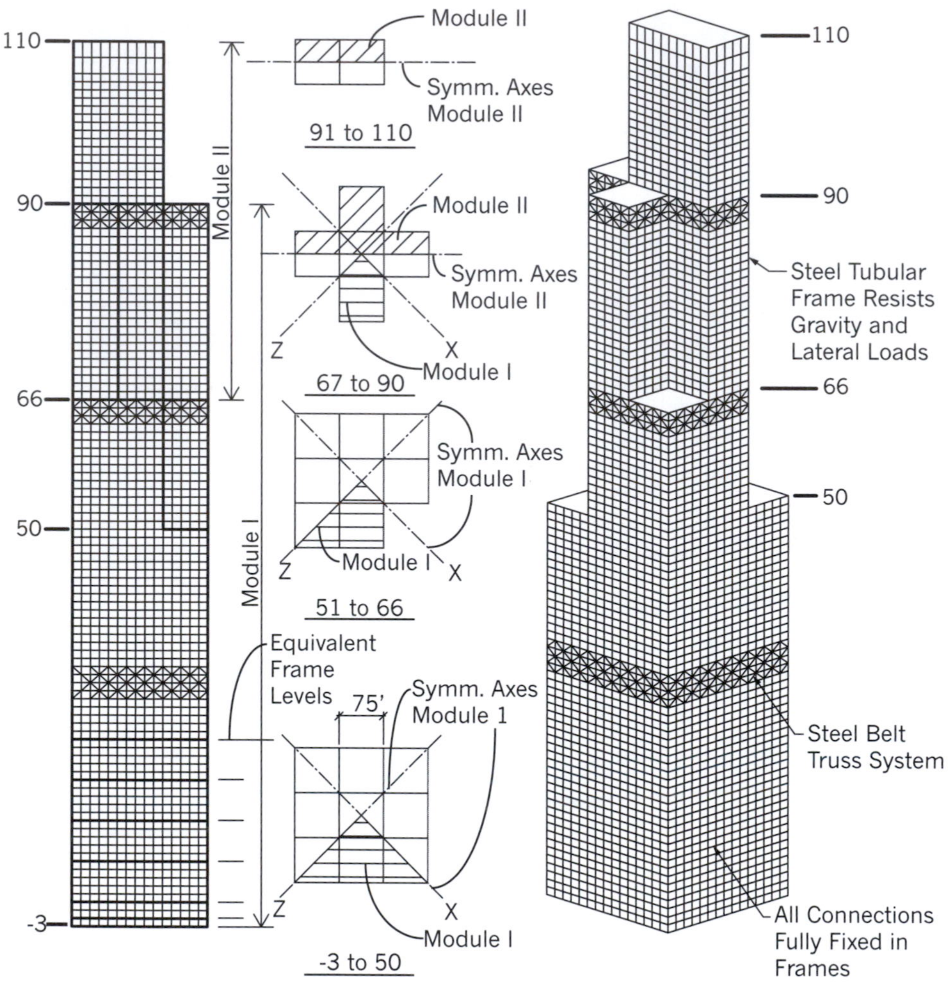

Lateral System Elevation, Willis Tower (formerly Sears Tower), Chicago, IL

Tustin Legacy Park, Tustin, California—Concentrically and eccentrically braced steel shear truss core combined with a perimeter steel frame resist lateral and gravity loads.

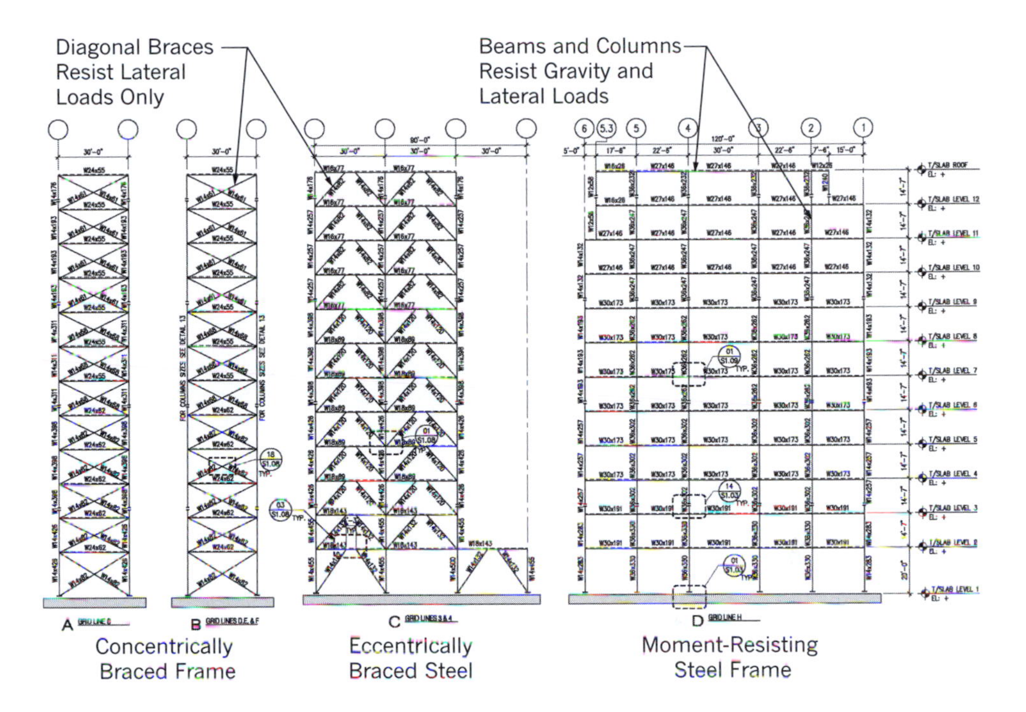

Lateral System Elevations, Tustin Legacy Park, Tustin, CA

5.3.2 Concrete

Gemdale Plaza, Beijing, China—Reinforced concrete superframe infilled with irregular screen frames on two of four façades and conventional reinforced concrete moment-resisting frames on the other two façades plus the central reinforced concrete shear wall core resist lateral and gravity loads.

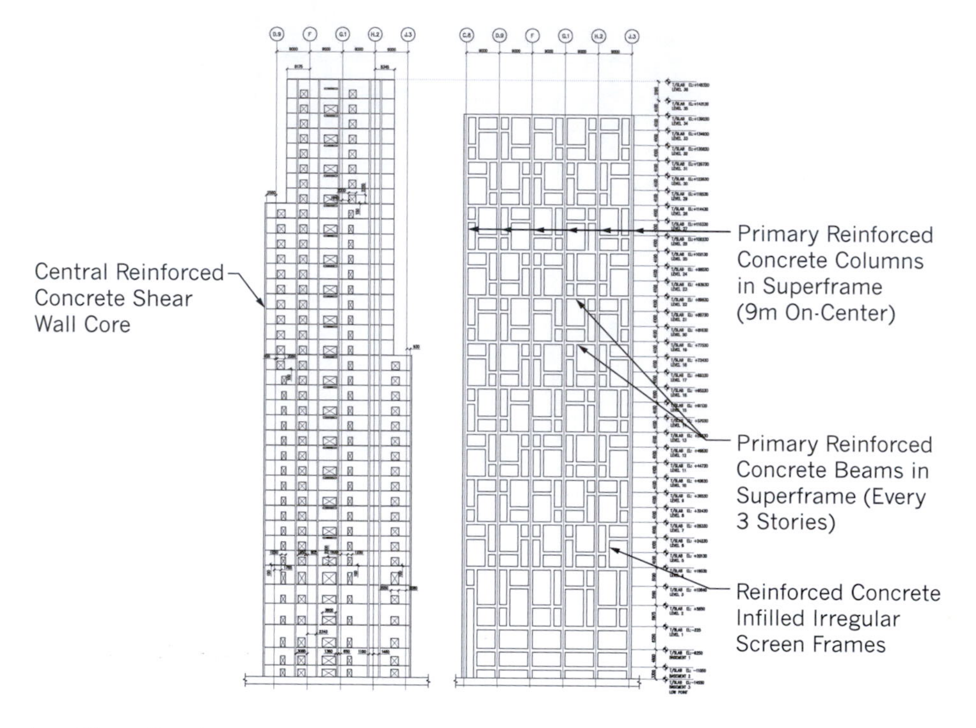

Lateral System Elevation,
Gemdale Plaza, Beijing, China

Burj Khalifa, Dubai, UAE—Central reinforced concrete shear wall buttressed core interconnected to perimeter reinforced concrete mega-columns resist both lateral and gravity loads.

Construction Image, Design Model, Burj Khalifa, Dubai, UAE

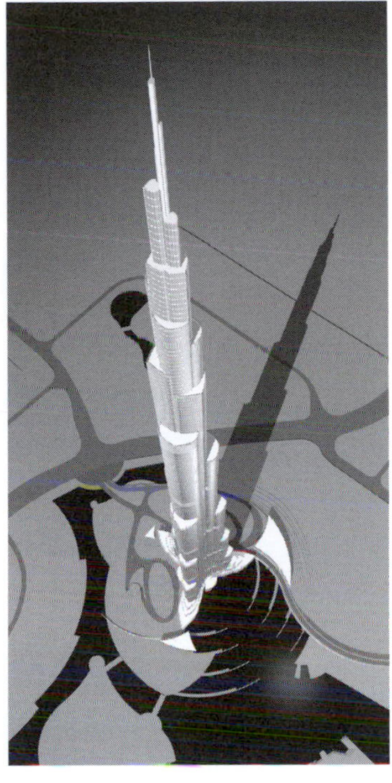

Burj Khalifa Design Model

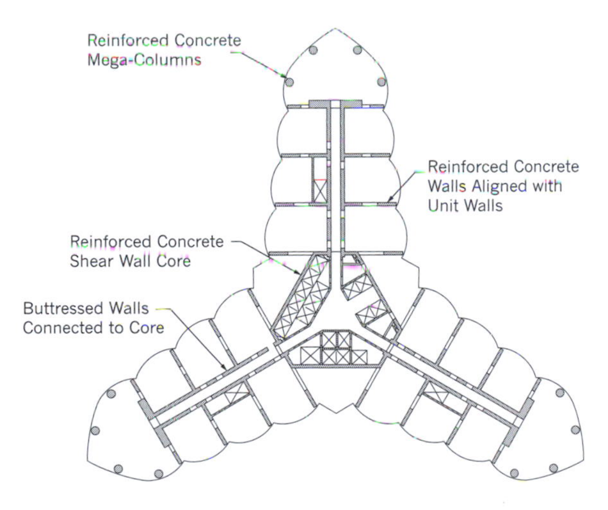

Floor Plan, Burj Khalifa, Dubai, UAE

5.3.3 Composite

Jin Mao Tower, Shanghai, China—A central reinforced concrete shear wall core is interconnected with composite mega-columns through outrigger trusses at three two-story levels. The central core and the perimeter mega-columns resist both gravity and lateral loads.

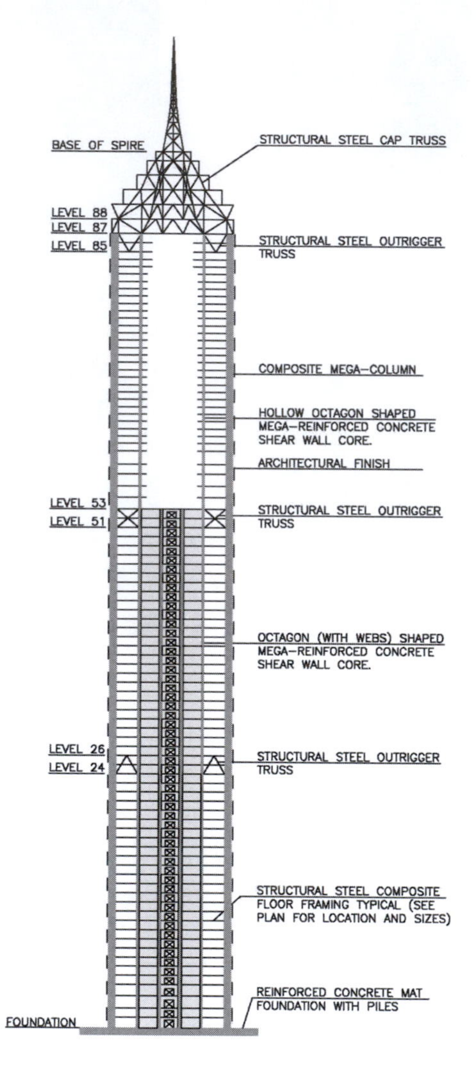

Lateral System Elevation, Jin Mao Tower, Shanghai, China

Tianjin Global Financial Center, Tianjin, China—A central steel-plated core (circular composite columns are interconnected by unstiffened steel plates) is interconnected with a perimeter composite moment resisting frame.

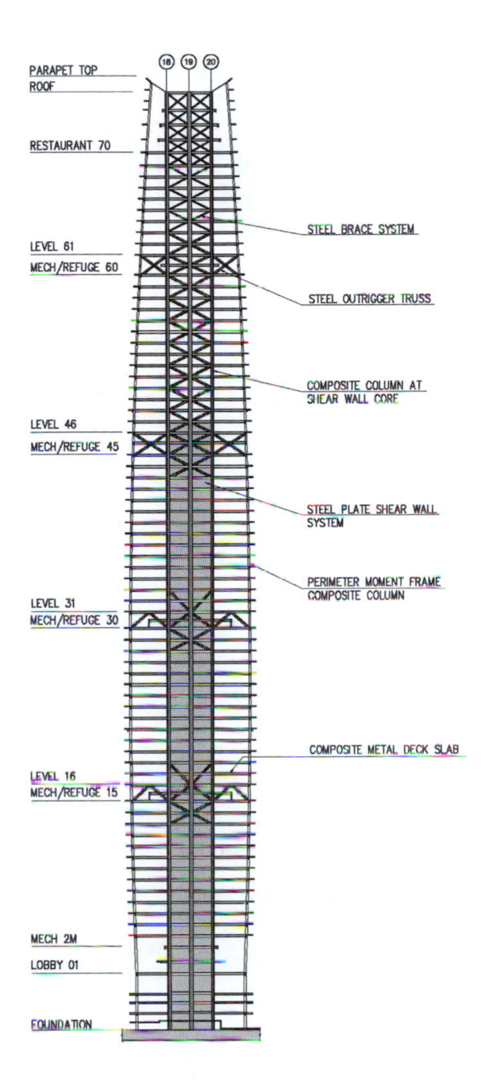

Lateral System Elevation, Tianjin Global Financial Center, Tianjin, China

CHAPTER 6

ATTRIBUTES

Strength including code limits and material types, and serviceability including drift, damping, wind-induced accelerations, creep, shrinkage, elastic shortening, have a fundamental effect on the design of the tower. The understanding of materials, building proportioning, and building behavior when subjected to loads are critical in determining both feasibility and successful use.

6.1 STRENGTH

Whether limit state (load and resistance factor design) or allowable stress design is used for members within the structural system, local codes and material type will dictate the method of design. Redundancy, load path, and the importance of structural elements within the system result in special structural design considerations. Strength design is usually based on wind loads with a return period of 50 years and seismicity with a 10% probability of exceedance in 50 years adjusted by a structural system-dependent reduction factor (response modification factor). The historic reliance on structural steel for tall towers has evolved into the use of ultra-high strength concrete with common production compressive strengths of 110 MPa (16,000 psi) and higher, compressive strengths approaching that of early cast irons/structural steels. Grade 36 steel has merged with Grade 50 steel and higher strength steels with yield strengths of 448 MPa (65 ksi) and higher are commonplace.

This increase in available strengths for concrete and steel has allowed more efficient designs with smaller structural elements. In addition, the combination of structural steel and reinforced concrete (composite) in structures has led to extremely efficient solutions.

FACING PAGE
Burj Khalifa Under
Construction, Dubai, UAE

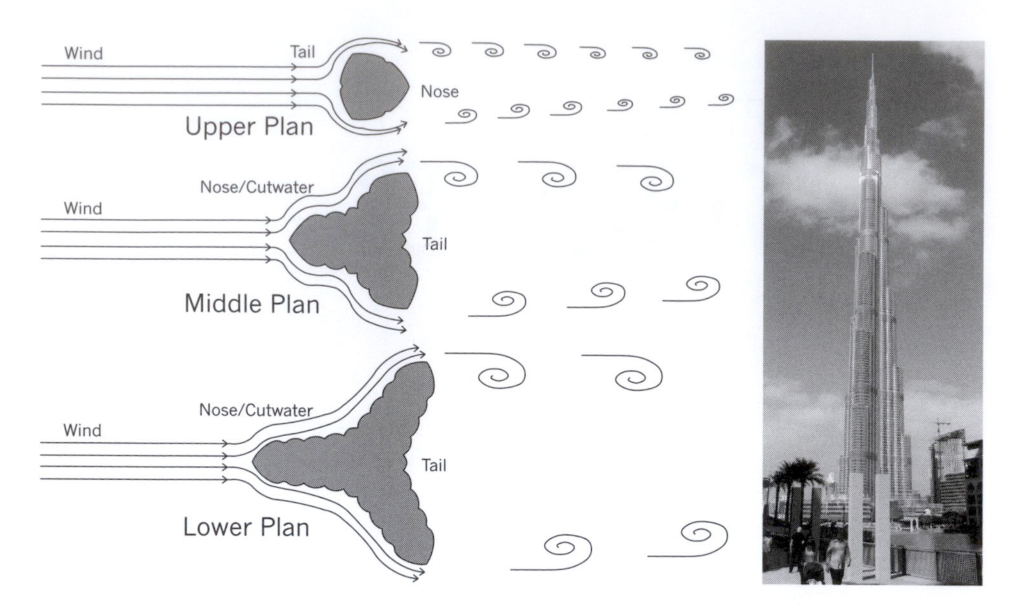

Disruption of Organized Vortex Shedding, Finished Photograph of Tower, Burj Khalifa, Dubai, UAE

6.2 SERVICEABILITY

In addition to strength considerations, serviceability of the tall tower is likely the most important design consideration and, at times, the least understood. Most consider drift to be the controlling factor for stiffness; in fact, building acceleration due to wind-induced motion can be far more critical. Evaluating occupants' perception to motion is based on building use, stiffness, mass, and damping of the structure.

6.2.1 Drift

The internationally recognized drift criterion for ultra-tall structures is h/500, typically based on elastic deformations (cracked sections in some reinforced concrete members including link beams and moment-frame beams) and a 50 year-return wind. Historically, some tall structures have been designed with allowable drifts as high as h/400.

In developing rational, applied wind pressure diagrams for the structure, specific damping ratios should be considered based on building materials and non-structural components (see following section for damping).

It is difficult to find codes that commit to the allowable building drifts for structures subjected to wind. The Canadian Building Code (h/500 for all structures) and the Chinese National Building Code are exceptions. On the

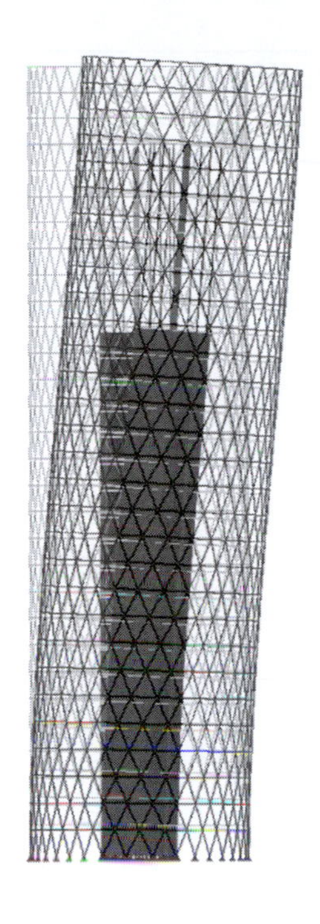

Deflection Shape Due to Seismic Loading, Building Under Construction, Agile Tower, Guangzhou, China

other hand, seismic drift limits are recognized by the UBC and are dependent on the structural system used. The limit is as follows:

$$\Delta m = 0.7\ R\ \Delta s$$

where:

Δm = maximum inelastic response drift (Δm shall not exceed 0.025h for T<0.7 seconds and 0.02h for T≥0.7 seconds, or h/40 and h/50 respectively)

R = lateral system coefficient representative of inherent overstrength and global ductility

Δs = maximum elastic drift

Some codes, such as China's National Building Code, take a strict position of allowable building drift for both wind and seismic conditions with system specific criteria. The limits are based on a 50 year-return wind, frequent earthquakes (62.5% probability of exceedance in 50 years), and elastic section properties (including gross section properties for reinforced concrete). The following chart summarizes major systems and limits.

Structural System	**Any Height**	**$H < 150m$**	**$150m \le H \le 250m$**	**$H > 250m$**
STEEL				
Wind (roof)	h/500	-	-	-
Wind (interstory)	h/400	-	-	-
Seismic (freq)	h/250	-	-	-
Seismic (rare)	h/70	-	-	-
CONCRETE (WIND AND SEISMIC FREQUENT)				
FRAME	-	h/550	interpolate	h/500
Frame-Shear wall (SW)	-	h/800	interpolate	h/500
SW Only, Tube-in-Tube	-	h/1000	interpolate	h/500
COMPOSITE (WIND AND SEISMIC FREQUENT)				
Steel Frame-SW	-	h/800	interpolate	h/500
Comp Frame-SW	-	h/800	interpolate	h/500
Comp Frame (steel beams)	h/400	-	-	-
Comp Frame (comp beams)	h/500	-	-	-
CONCRETE FRAME (SEISMIC RARE)				
Frame	h/50	-	-	-
Frame - SW	h/100	-	-	-
SW - Tube-in-Tube	h/120	-	-	-
COMPOSITE (SEISMIC RARE)				
Comp Frame	h/50	-	-	-
Other	h/100	-	-	-

Second order (P-Δ) effects must be considered in tall structures laterally displaced by wind and seismic loads. These effects could increase drift by as much as 10% and must be also considered for the strength design of lateral load resisting members. Examples for building drifts for major building projects are listed as follows:

Building	**Height**	**Drift**	**Material**
Sears Tower (Chicago)	445 m	H/550	Steel
Jin Mao Tower (Shanghai)	421 m	H/908	Mixed
Central Plaza (Hong Kong)	374 m	H/780	Concrete
Amoco Building (Chicago)	346 m	H/400	Steel
John Hancock (Chicago)	344 m	H/500	Steel
Columbia Seafirst (Seattle)	288 m	H/600	Mixed
Citibank Plaza (Hong Kong)	220 m	H/600	Mixed

6.2.2 Damping

Damping of the tall towers can have a significant effect on design forces and wind-induced accelerations. Damping is material specific and is proportional to demand on the structure. Theoretical analysis, laboratory testing, and in situ monitoring have provided general requirements for design. The overall building damping typically considered in the behavior of tall building structures is as shown in the table below.

MATERIAL	Return Period for Wind			
	1-10 year	50 year	100 year	1,000 year (collapse prev.)
Concrete	2%	3%	5%	7%
Steel	1%	2%	3%	4%

The building damping values may be specifically calculated as follows:

$$\xi = \xi_N + \xi_M + \xi_{SD} + \xi_{AE} + \xi_{SDS}$$

Where:

ξ = total building damping ratio

ξ_N = non-structural component damping (1–1.5%)

ξ_M = material damping (concrete uncracked members = 0.75%, Steel = 0%)

ξ_{SD} = structural damping (concrete cracked members = 0.5–1.5%, Steel = 0–0.5%)

ξ_{AE} = aero-elastic damping (0–0.75%)

ξ_{SDS} = supplemental damping systems (viscoelastic = 5–30%, vibration absorbers = 1–5%)

And:

$$\xi_{SD} = \frac{\sum E_D}{4\pi E_{SO}}$$

Where:

E_D = energy loss per element per full cycle to a given performance level

E_{SO} = total building elastic strain energy associated with a given performance level

$$E_{SO} = \sum \frac{1}{2} F_i \Delta_i$$

Where:

F_i = wind force at each level for a given direction

Δ_i = corresponding displacement at the point of loading, at each level

$$\sum E_D = k * \sum E_{SD}$$

Where:

k = adjustment factor, $k = \frac{4\pi (E_{SOmodel}) \xi_{SD\text{"}measured\text{"}}}{\sum E_{SDmodel}}$

E_D = total energy loss in the members ("measured")

E_{SD} = total strain energy in the members (model)

6.2.3 Accelerations

Without acceptable levels of wind-induced accelerations, the tall tower can be unusable during strong wind events. Across-wind accelerations or lift accelerations are usually more serious than along-wind or drag accelerations. There have been recorded instances where occupants of super-tall buildings have perceived building motion, felt nauseous, and in some cases left the building during a windstorm. Other visual or audible conditions during windstorms lead to discomfort. Perceived motion relative to neighboring structures is most acute especially due to torsion. Water in toilets may slosh. Exterior wall elements or interior partitions may creak. Wind speed, building height, orientation, shape, and regularity along the elevation all contribute to the behavior.

Apartment in the John Hancock Center, Chicago, IL

The inner ear is very sensitive to motion. A person lying down is more susceptible than one in a seated or standing position. People who reside in structures (apartments or condominiums) rather than occupy them transiently (office buildings) are generally more susceptible to perceiving building movement. The limits of perceptible acceleration are:

Occupancy Type	Horizontal Accelerations Return Wind Period	
	1 year	10 year
Office	10-13 milli-g's	20-25 milli-g's
Hotel	7-10 milli-g's	15-20 milli-g's
Apartment	5-7 milli-g's	12-15 milli-g's

These accelerations are usually most critical at the top occupied floor and are calculated for building damping ratios of 1% of critical steel to 1½% of critical for composite or reinforced concrete structural systems. The return period refers to the maximum accelerations statistically expected for maximum winds over a defined period of time.

Torsional accelerations/velocities experienced by structure are in many cases more important than horizontal accelerations. This is especially true where occupants have a point of reference relative to neighboring structures. The limit of acceptable torsional velocity is 3.0 milli-radians/sec. The National Building Code of Canada Structural Commentary Part 4 offers a calculation method to predict horizontal accelerations. This method provides an excellent preliminary calculation that can be later confirmed by the rational wind tunnel studies.

Across-wind accelerations are likely to exceed along-wind accelerations if the building is slender about both axes, that is if:

$$\sqrt{WD}\,/\,H < \tfrac{1}{3}$$

where:

W = across-wind plan dimension (m)
D = along-wind plan dimension (m)
H = height of the building (m)

For these tall, slender structures, accelerations caused by wind-induced motion are defined as:

$$a_w = n_w{}^2 g_p \sqrt{WD}\left(\frac{a_r}{\rho_B g \sqrt{\beta_w}}\right)$$

For less slender structures or for lower wind speed, the maximum acceleration is:

$$a_D = 4\pi^2 n_D^2 g_p \sqrt{\frac{K_S F}{C_e \beta_D}} \frac{\Delta}{C_g}$$

where:

a_w, a_D = peak acceleration in across-wind and along-wind directions, m/s^2

a_r = $78.5x10^{-3}\left[V_H / \left(n_w \sqrt{WD}\right)\right]^3$, Pa

ρ_B = average density of the building, kg / m^3

β_w, β_D = fraction of critical damping in across-wind and along-wind directions

n_w, n_D = fundamental natural frequencies in across-wind and along-wind directions, Hz

Δ = maximum wind-induced lateral deflection at the top of the building along-wind direction, m

g = acceleration due to gravity = $9.81 m/s^2$

g_p = a statistical peak factor for the loading effect

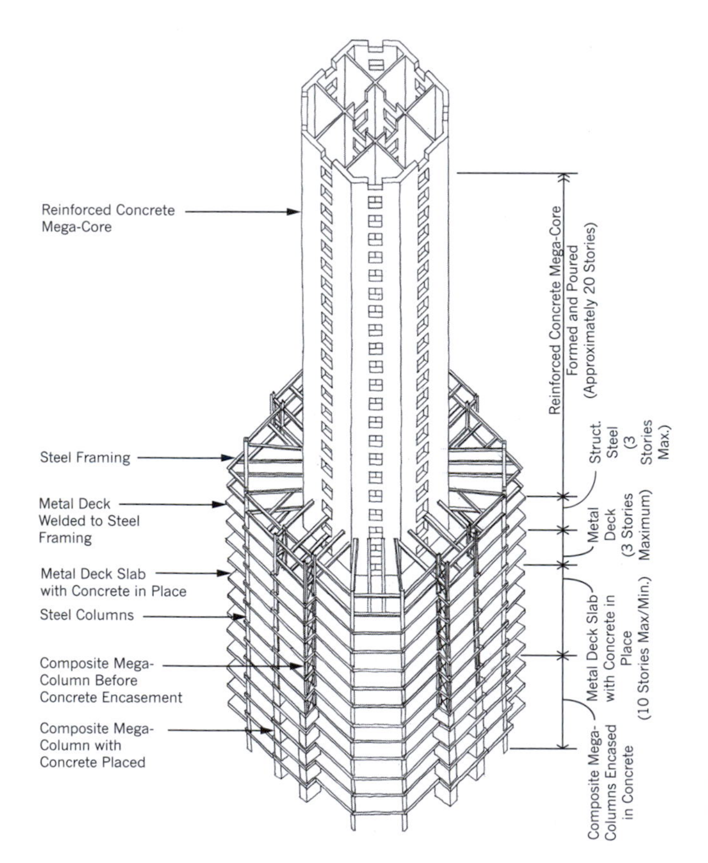

Mixed Use of Structural Materials/Construction Sequence, Jin Mao Tower, Shanghai, China

K = a factor related to the surface roughness coefficient of the terrain
s = size reduction factor
F = gust energy ratio
C_e = exposure factor at the top of the building
C_g = dynamic gust factor

Supplemental damping may be introduced into the structure to control accelerations. This damping may consist of tuned mass, sloshing, pendulum, or viscous damping systems.

6.2.4 Creep, Shrinkage, and Elastic Shortening

Vertical elements within the floor plan of tall towers are subjected to shortening. This shortening begins during the construction process and can continue up to 10,000 days after construction is complete. Vertical displacements affect non-structural components, including the exterior wall, interior partitions, and vertical MEP (mechanical/electrical/plumbing) systems.

Construction Image,
Jin Mao Tower, Shanghai, China

Construction Image, Creep, Shrinkage, and Elastic Shortening Results, Jin Mao Tower, Shanghai, China

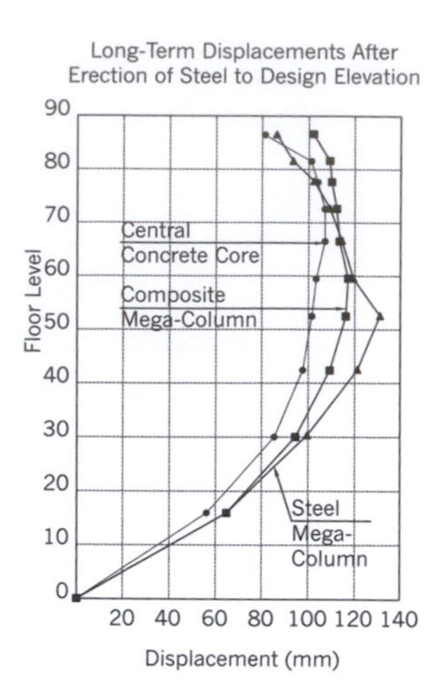

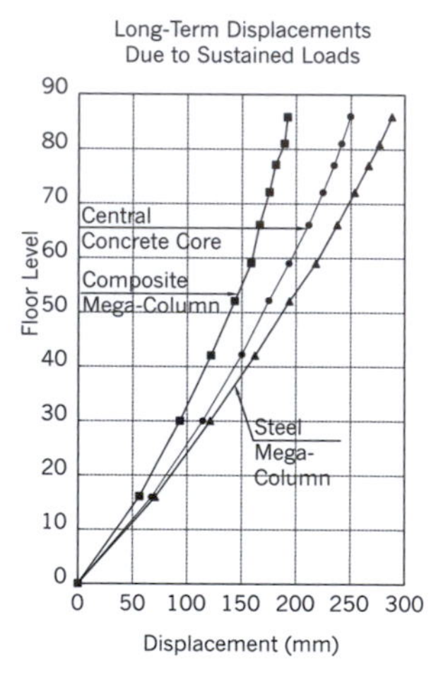

Construction to Design Elevation, Jin Mao Tower, Shanghai, China

These displacements are known to cause failure in supports for water piping within tall structures where creep and shrinkage particularly in concrete towers occur long after the pipes have been placed. Vertical displacements of 300 mm (12 in) or more at the top of the structure are not uncommon. Relative displacements between vertical elements on a floor plate can cause sloped floors or overloading of interconnection elements not designed for the relative movement.

Steel structures are more easily predictable when evaluating long-term relative movement between vertical elements within the structural system. Their displacements are only affected by elastic shortening due to axial load, assuming that the vertical elements are not permanently affected by eccentric long-term loads. Therefore, the variables in considering relative displacement are the self-weight dead loads and their distribution to vertical elements, and superimposed dead and live loads after the structure is complete.

Relative displacements between vertical elements in reinforced concrete or composite structures are far more difficult to predict since time, geometry, material composition, curing, and load all contribute to short and long-term creep, shrinkage, and elastic shortening. Equalization of applied stress to these elements is an important design consideration. For preliminary calculations, a total strain of 700×10^{-6} in/in can be used.

The following guidelines should be used in designing for creep, shrinkage, and elastic shortening:

1. Determine construction sequence.
2. Calculate design loads (only realistic, sustained loads must be considered) and attempt to consider the actual compressive strength anticipated for the structure (in most cases, the actual in-place concrete compressive strength could be 10–25% higher than the theoretical design value). Over 90% of the sustained load in central reinforced shear wall system is likely to be dead plus superimposed dead load. Of the total load on an exterior steel column, 75% is likely to be dead load plus superimposed dead load, 15% exterior wall load, and 10% live load.
3. Calculate anticipated vertical displacements.
4. Develop a construction program that requires the contractor to build the structure to "design elevation." This can be done with laser surveying techniques and making modifications to structural elements during construction (i.e. adjusting formwork heights, providing prefabricated shims for structural steel).
5. Establish displacements at the time of construction of elements at a particular floor elevation. This method allows for "zeroing out" of displacements within the system to this point.
6. Determine corrections that are required to control relative displacements between elements. It is important not to overcompensate, remembering that an element placed higher than the design elevation must "travel" through a theoretical zero or level point and can be located below a relative member and still be within acceptable tolerances.

Elastic Strain Equation at time *i*:

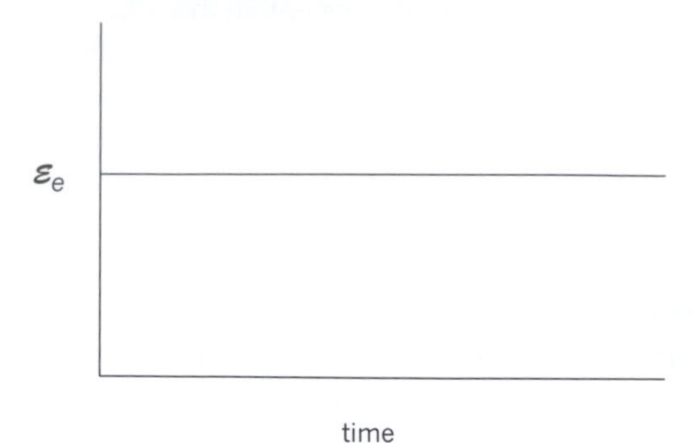

$$\varepsilon e_i = \frac{P_{g_i}}{A_t E_{c_i}} + \varepsilon e_{i=1}$$

where:

ε_{e_i} = total elastic strain at time i
P_{g_i} = incremental gravity load applied at time i (kN)
A_{t_i} = transformed section area at time i (mm^2)
E_{c_i} = concrete modulus of elasticity at time i (MPa)
$\varepsilon_{e_{i=1}}$ = total elastic strain at the previous time interval

Shrinkage Strain Equation at time *i*:

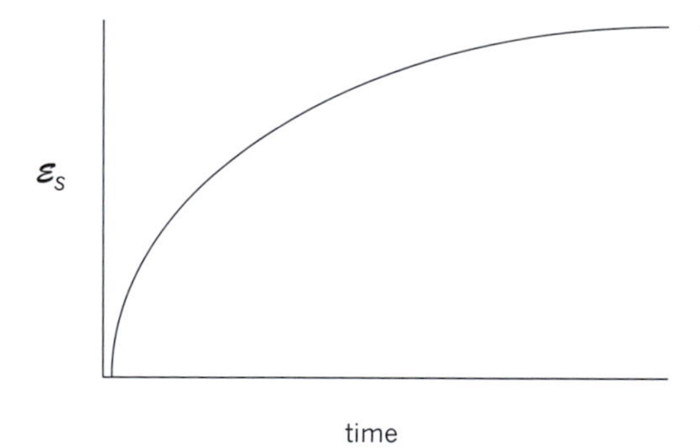

$$\varepsilon_{s_i} = \left(\varepsilon_{s_i w} - \varepsilon_{s_{i-1} w}\right) R_{cf_i} + \varepsilon_{s_{i-1}}$$

where:

ε_{s_i} = total shrinkage strain at time i
$\varepsilon_{s_i w}$ = baseline shrinkage strain at time i
$\varepsilon_{s_{i-1} w}$ = baseline shrinkage strain at the previous time interval
R_{cf_i} = steel reinforcement correction factor at time i
$\varepsilon_{s_{i-1}}$ = total shrinkage strain at the previous time interval

Incremental Creep Strain Equation at time *i* due to load at time *j*:

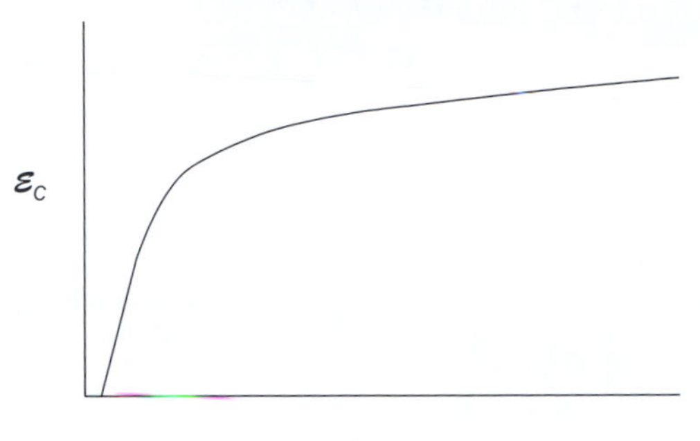

time

$$_{c(i-1)\rightarrow ij} = \left(\varepsilon_{c_i w_j} - \varepsilon_{c_{(i-1)} w_j}\right) R_{cf_i}$$

where:

$_{c(i-1)\rightarrow ij}$ = incremental creep strain at time *i* due to load at time *j*

$\varepsilon_{c_i w_j}$ = creep strain at time *i* due to load at time *j* without rebar effect

$\varepsilon_{c_{(i-1)} w_j}$ = creep strain at the previous time interval due to load at time *j* without rebar effect

R_{cf_i} = steel reinforcement correction factor at time *i*

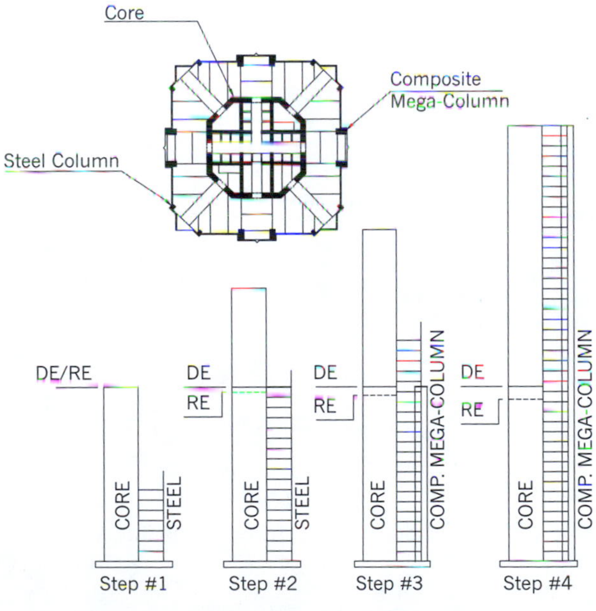

DE = Building Design Elevation
RE = Reference Elevation

Construction to Design Elevation

The Plaza
Roof Garden

CHAPTER 7
CHARACTERISTICS

DYNAMIC PROPERTIES, aerodynamics, placement of structural materials, floor-to-floor heights, and aspect ratios are all important characteristics in maximizing structural efficiency of the tall tower.

7.1 DYNAMIC PROPERTIES

The fundamental period of the tall tower roughly can be estimated by considering the number of anticipated stories divided by 10. According to the ANSI (ASCE-88) Code, the following periods may be calculated for translational and torsional responses of tall structures:

Steel buildings: $T = 0.085H^{0.75}$
Concrete buildings: $T = 0.061H^{0.75}$
Steel or concrete buildings: $T_\theta = 0.054N$

Where:

T = fundamental translational period
T_θ = first torsional period
H = building height in meters
N = number of stories

Therefore, using the above equations for a 50-story tower with an average floor-to-floor height = 4 m (13 ft, 2 in):

$$T_{steel} = 0.085\ (200)^{0.75} = 4.5 \text{ seconds}$$
$$T_{concrete} = 0.061\ (200)^{0.75} = 3.2 \text{ seconds}$$

FACING PAGE
Al Hamra Tower Under Construction, Kuwait City, Kuwait

Building (HT, Material)	Fundamental Trans. Modes	Torsional Mode
Jin Mao Tower (H = 421 m, mixed)	T_1 = 5.7s, T_2 = 5.7s	T_θ = 2.5s
Burj Khalifa Tower (H = 828 m, r/c)	T_1= 11.0s, T_2 =10.0s	T_θ = 4.0s
Al Hamra Tower (H = 415 m, r/c)	T_1= 7.5s, T_2= 5.9s	T_θ = 3.2s
Gemdale Plaza Tower (H = 150 m, r/c)	T_1 = 4.4s, T_2 = 3.6s	T_θ = 2.4s
Tianjin Global Financial Center (H = 339 m, steel)	T_1= 8.2s, T_2= 7.5s	T_θ = 6.1s
Kingtown International Center (H = 235 m, mixed)	T_1= 5.0s, T_2= 4.8s	T_θ = 3.6s

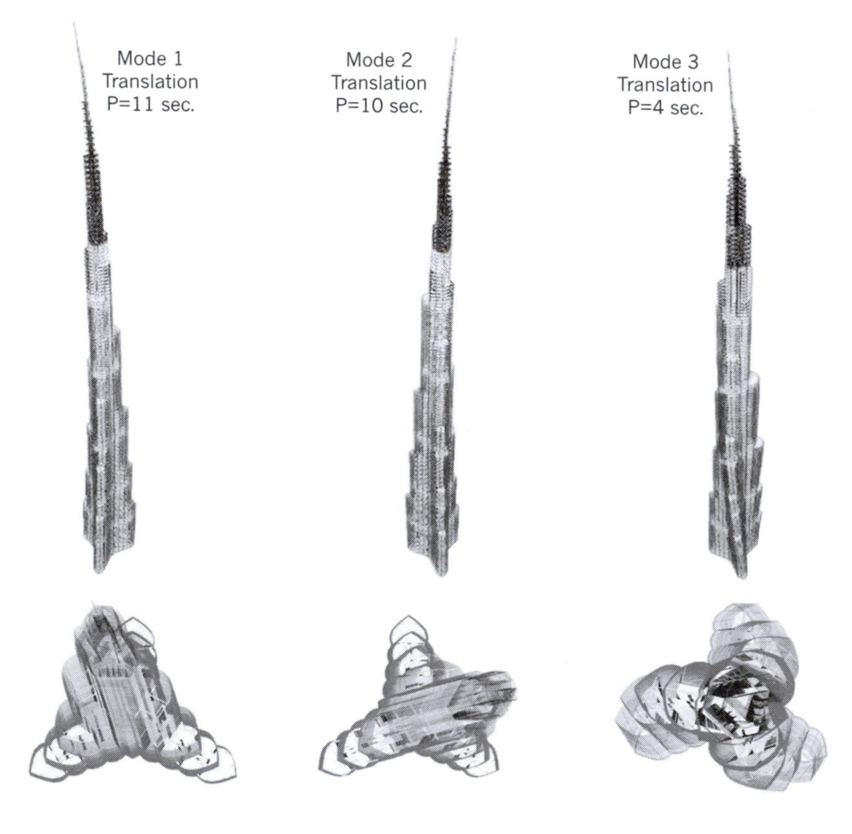

Dynamic Behavior,
Burj Khalifa, Dubai, UAE

7.2 AERODYNAMICS

The aerodynamics of the tall tower are important in minimizing imposed design forces. Across-wind motion (perpendicular to the direction of applied wind load) governs the behavior of the structure. Organized vortex shedding generates the highest forces levels. Vortex shedding is most organized (has the most adverse effects) with circular cross-sectional shapes, less organized with triangular shapes, and least organized with square shapes. The introduction of holes through the building cross-sections improves the behavior further. Variation of the structure's cross-section along its height also acts to disorganize or disseminate vortex shedding. Strouhal numbers describe oscillating flow mechanisms and are based on the frequency of vortex shedding, shape geometry, and wind velocity. Strouhal numbers for common building shapes are shown below in parentheses.

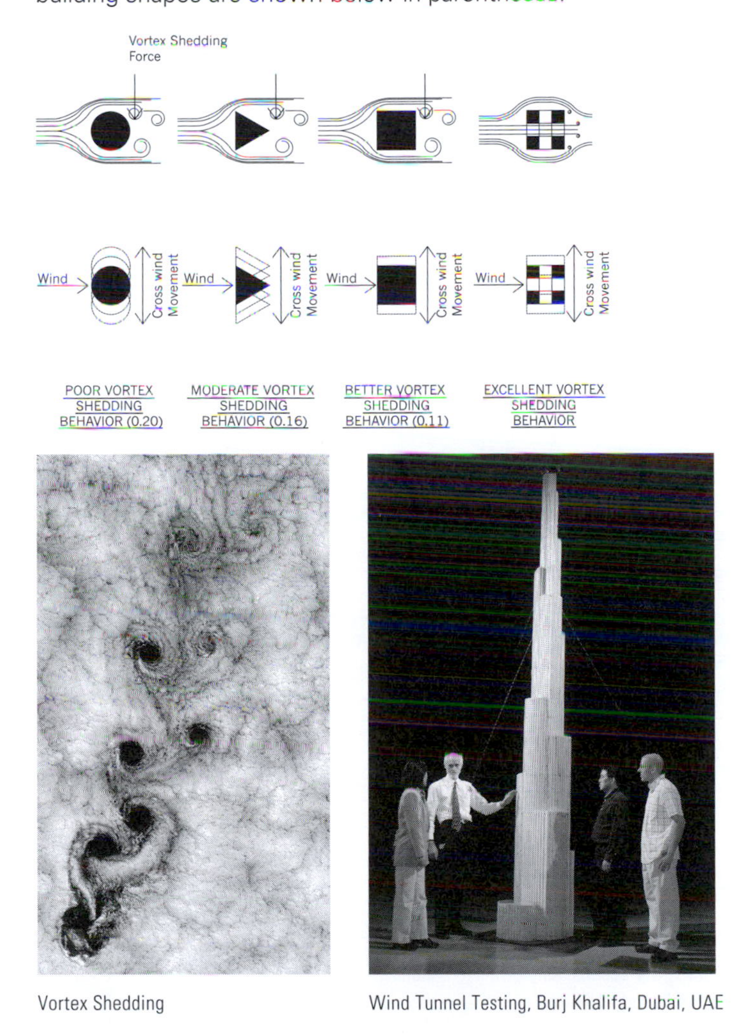

Vortex Shedding

Wind Tunnel Testing, Burj Khalifa, Dubai, UAE

7.3 PLACEMENT OF STRUCTURAL MATERIALS

Placement of structural materials within the tall tower is crucial to efficiency and economy. Placing material at the perimeter of the structure leads to the highest effective stiffness. Large material concentrations (columns) at the four (4) corners of a square plan are most efficient with only 50% efficiency realized with concentrated materials at the midpoint of the face of a square or distributed evenly along a circular plan.

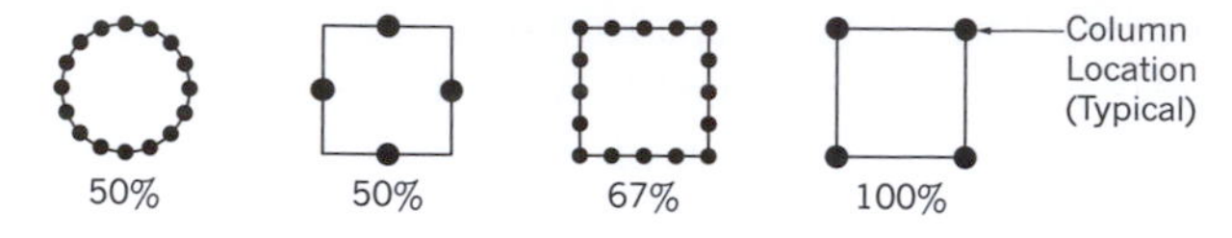

7.4 FLOOR-TO-FLOOR HEIGHTS

Minimizing floor-to-floor heights maximizes use within the structure. Every effort should be made to coordinate building systems and clear ceiling heights. The desired number of floors can reduce the building height, or the number of floors used within a specified height limit can be maximized. For office use, a 9 ft tall ceiling height nominally yields a floor-to-floor height of 13 ft, 1.5 in, and for residential use a nominal floor-to-floor height of 10 ft, 6 in

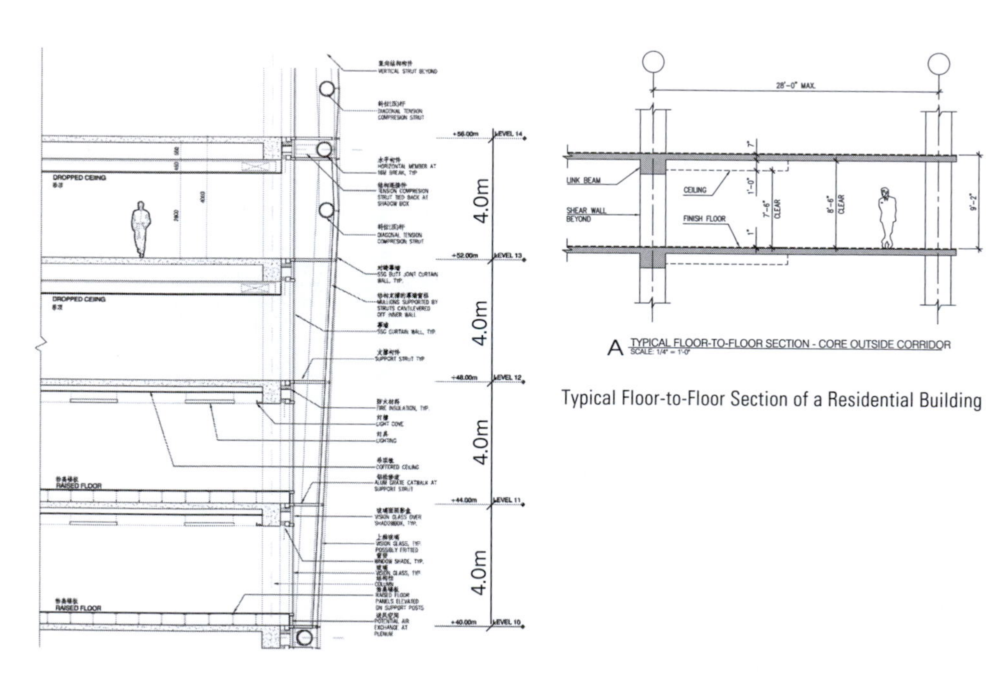

Typical Floor-to-Floor Tower Section, Kingtown International Center (formerly Jinao Tower), Nanjing, China

Typical Floor-to-Floor Section of a Residential Building

can be used for the basis of design with a minimum finished ceiling height of 8 ft (could be considerably higher without drop ceiling).

7.5 ASPECT RATIOS

The aspect ratio of a building is the ratio of height to the least structural width of the building. For super-tall buildings, it is essential to activate the full structural width of the building. With structural systems located at the perimeter of the building, a target aspect ratio between 6 and 7 to 1 is common. For tall towers, this ratio may be 8 to 1 or more. Typically, buildings with aspect ratios greater than 8 to 1 should consider supplemental damping systems to reduce occupant perception to motion. Shear wall cores centrally located typically have aspect ratios of between 10 and 15 to 1.

Building	Height	Aspect Ratio (height/width)	Material
Willis Tower (formerly Sears Tower)	445 m	6.4	Steel
Jin Mao Tower	421 m	7.0	Mixed
Amoco Building	346 m	6.0	Steel
John Hancock	344 m	6.6	Steel

C = properties of structure; H = height; D = least footprint dimension (width)

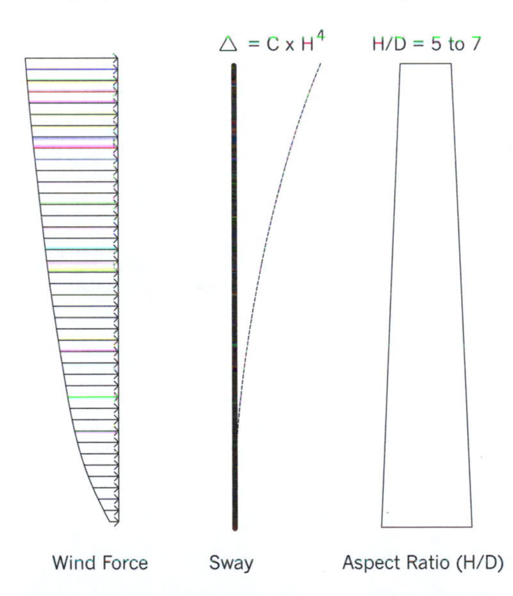

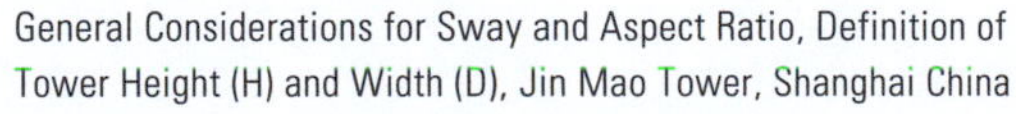

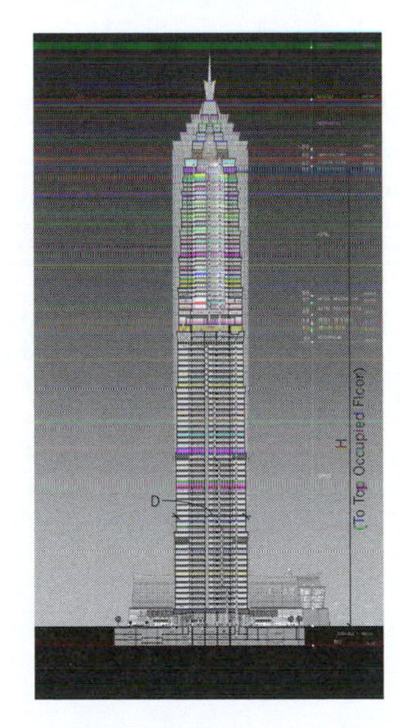

General Considerations for Sway and Aspect Ratio, Definition of Tower Height (H) and Width (D), Jin Mao Tower, Shanghai China

CHAPTER 8

SYSTEMS

THERE ARE SEVERAL FACTORS to consider when selecting a structural system for tall buildings. Safety, occupant comfort, and economy are the most important. The use and aesthetic of the structure also dictate possible solutions. For instance, it may be desirable to have wide column spacing at the perimeter rather than closely spaced columns. Available construction materials, available construction time, and contractor expertise also must be considered. Site conditions such as poor soil conditions or high seismicity may prescribe a particular structural system.

Building use is an important consideration when selecting a structural system. For instance, residential (or hotel) construction generally aligns well with a reinforced concrete structure. This structural system allows for smaller spans and in many cases flat plate (slab) floor systems. Post-tensioning can be used to increase spans while maintaining a minimal structural framing depth. Generally, a 9 m x 9 m (30 ft x 30 ft) column spacing works well with residential use and can be readily coordinated with parking requirements that typically exist below the residential structure. The underside of this framing system can be finished and painted serving as an acceptable finished ceiling while maximizing usable height for the spaces. A structural steel or composite system is also appropriate for this type of construction, and may be best if construction speed is an issue, but requires fireproofing and finished ceilings.

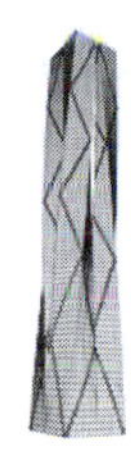

Structural System Sketches,
Jinling Tower Design Concept, Nanjing, China

FACING PAGE
Jinling Tower Design Concept
Rendering, Nanjing, China

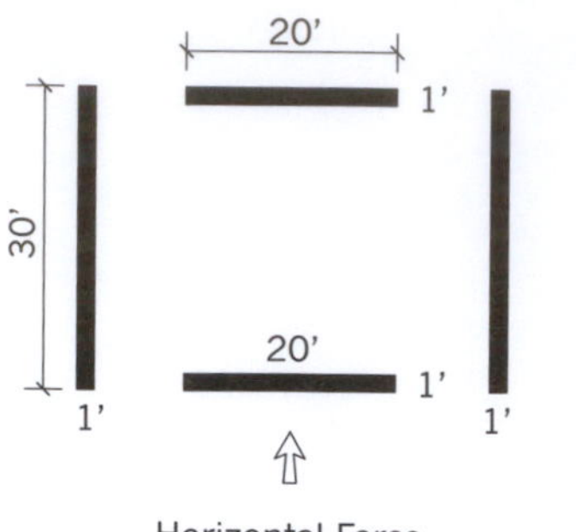

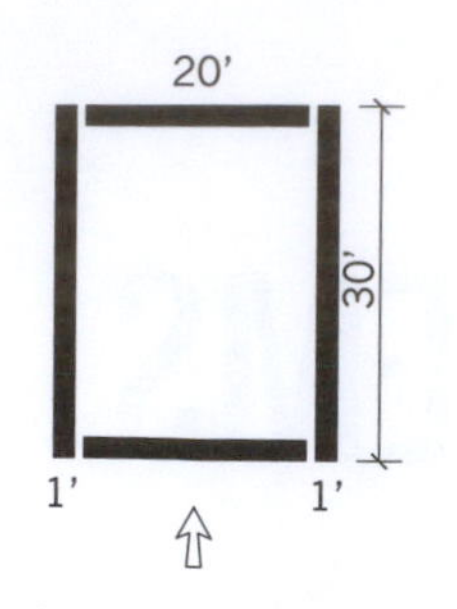

Moment of Inertia Comparison, Structural System in Plan

Reinforced concrete systems can also be used for office buildings, but long-span conditions should be carefully considered. Ideal office space configurations result in spans of approximately 13.5 m (45 ft), resulting in rather deep conventional reinforced concrete beam framing. Post-tensioning can be used in the beam framing to reduce overall depth. It is generally not advised to use post-tensioning in floor slabs for an office building since tenant modifications (interconnecting stairs etc.) over the life of the structure are difficult to incorporate since post-tensioning tendons cannot be cut. Designated areas where conventional reinforcing could be used and raised floors (services are distributed through raised floor rather than through structure) generally alleviate these limitations.

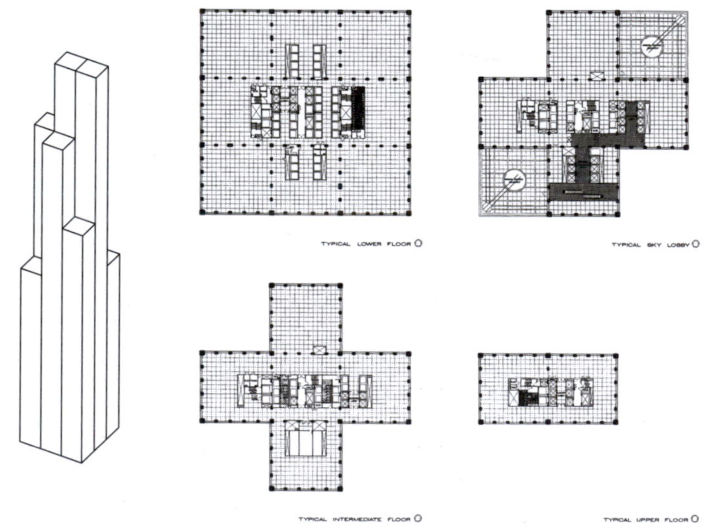

Tubular Structural System Concept and Resulting Floor Plans, Willis Tower (formerly Sears Tower), Chicago, IL

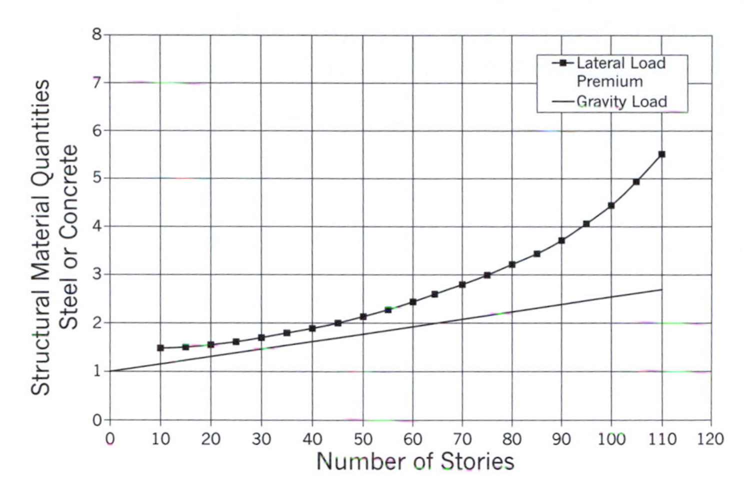

Structural Material Quantities vs. Number of Stories

Structural steel or composite systems can be used for either residential or office construction. For residential uses, the floor-to-floor heights generally increase (finished ceilings required underneath steel framing systems) and typically ceilings are coffered around framing to give maximum ceiling heights. For office buildings, steel framing works well, allowing for future modifications (local areas generally can be reframed to accommodate openings, etc.) and long-span conditions can be readily addressed. Wide-flanged framing with mechanical systems distributed below the beams, or built-up truss framing with mechanical systems distributed through the trusses, can be used. Framing is fire-proofed with finished ceilings generally required. Raised floor systems can be incorporated to allow for mechanical air distributions, as well as electrical and data distributions.

8.1 MATERIAL QUANTITIES

In general, structural materials required to resist gravity loads are fairly constant for low-to-mid-rise structures, but increase linearly for taller structures. In addition, the amount of material required for the lateral load-resisting system increases significantly with height.

A building is usually considered a "high-rise" if its height exceeds 23 m (75 ft). At extreme heights of 610 m (2000 ft) and above, increasingly complex design considerations must be taken into account.

Generally, the minimum amount of material required for gravity loads is as follows (apply quantities to gross framed building area):

Steel building:	49 kg/m^2 (10 psf)
	0.1 m^3/m^2 (0.4 cu ft/sf) (concrete)*
	4.8 kg/m^2 (1.0 psf) (rebar)*
Reinforced concrete building:	32 kg/m^2 (6.5 psf) (rebar)
	0.34 m^3/m^2 (1.1 cu ft/sf) (concrete)

*Concrete and rebar required typically for metal deck slabs systems in steel buildings.

As heights increase, considerations for structural systems change. For instance, the steel frame used in the Reliance Building would require approximately 68 kg/m^2 (14 psf) of structural steel for the 15-story building height of 61.6 m (202 ft). If this same building frame were used in a building of 50 stories (approx. 200 m or 656 ft), the material required for strength and stiffness of the moment-resisting frame may be 244 kg/m^2 (50 psf) or more. On the other hand, if a braced frame were used in the core area and a steel frame at the perimeter, the material quantities required may be reduced to 122 kg/m^2 (25 psf).

8.2 PRACTICAL LIMITS OF STRUCTURAL SYSTEMS

As the number of stories and structural heights increase, structural solutions must respond to behavior typically controlled by lateral wind loads (and gravity). In regions of high seismicity, ductile detailing of the system components is crucial to successful performance. In many cases, lateral loads imposed by seismic ground motions can govern the design (strength or serviceability) based on mass and geometric and stiffness characteristics.

The John Hancock Center . . . was designed in six months' less time by computers.

X-Braces Trim Steel Tonnage

The Genius of Dr. Fazlur Kahn—Innovative Structural Systems and the Use of Main Frame Computing

When structural systems are defined for a tower structure, it is usually implied that the system describes to the lateral load resisting portion of the structure. The lateral system typically serves dual roles, resisting wind and seismic forces as well as resisting gravity loads. Many of these structures have additional structural elements that resist gravity loads only, including floor slabs, floor framing beams, and columns that support framing with shear or pin connections only.

Limits of structural systems based on the number of stories, and therefore height, are subjective, but based on years of development have proven to lead to the most efficient and safe solutions. For instance, a

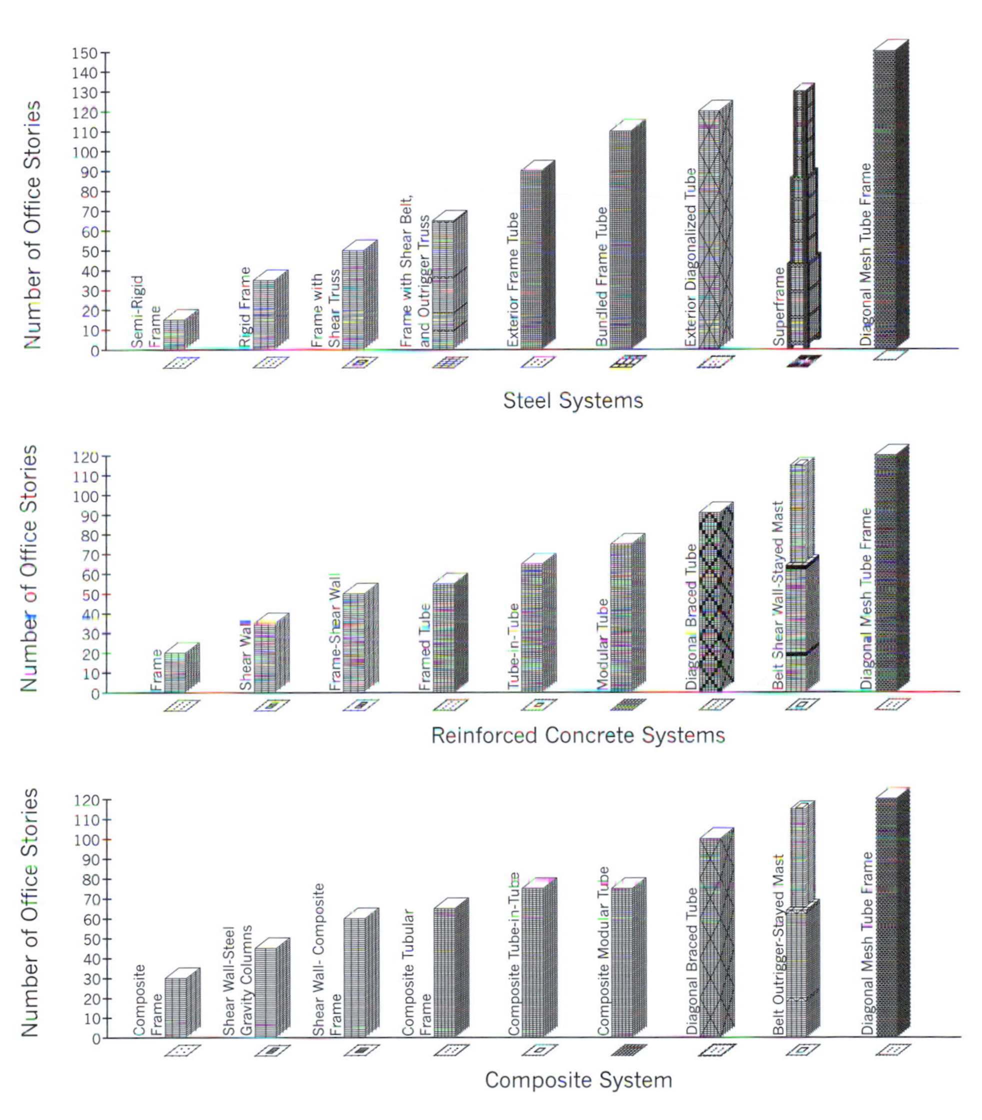

Practical Height Limits of Structural Systems

moment-resisting rigid frame may have a practical height limit of 35 stories or 140 m (460 ft), but can and has been used in structures of greater height (the Empire State Building is a good example of where a rigid moment frame was used for the 102 story, 382 m (1252 ft) tall structure). In each case the proposed structural solution must be superimposed and coordinated with the proposed architecture.

Practical limits of structural systems are shown in the following diagrams. The systems are typically associated with office floors based on the associated floor-to-floor heights, but can similarly be applied to other building uses.

8.2.1 Structural Steel

Building heights listed below are based on an average floor-to-floor height of 4.0 m (13 ft, 1½ in) allowing for a 2.75 m (9 ft) tall ceiling. Six meters (20 ft) are included for lobby spaces. One 8 meter (26 ft) tall space is included in the mid-height of the tower for mechanical spaces or a sky lobby for buildings 60 stories and above.

8.2.1.1 Steel Semi-Rigid Frame

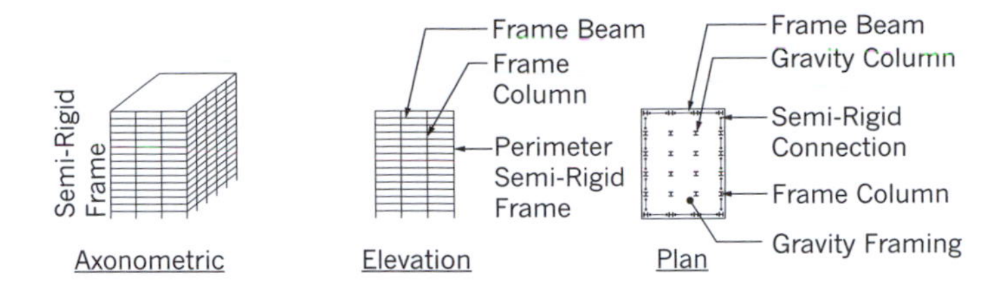

General limit of number of stories = 15 and height = 62 meters (203 ft, 4 in)

The moment-resisting frame typically comprises wide-flanged shaped columns and beams with primary frame connections typically bolted with slip-critical bolts. Frame connections allow for partial fixity allowing for some rotation when loaded. Column spacing typically ranges from approximately the floor-to-floor height to twice the floor-to-floor height; generally, spacing varies between 4.5 m (15 ft) and 9 m (30 ft). The taller the building, the greater the need to use deeper wide-flanged sections for beams and columns and closer column spacing. Columns typically vary from W14 to W36 sections and beams typically vary from W21 to W36 sections.

8.2.1.2 Steel Rigid Frame

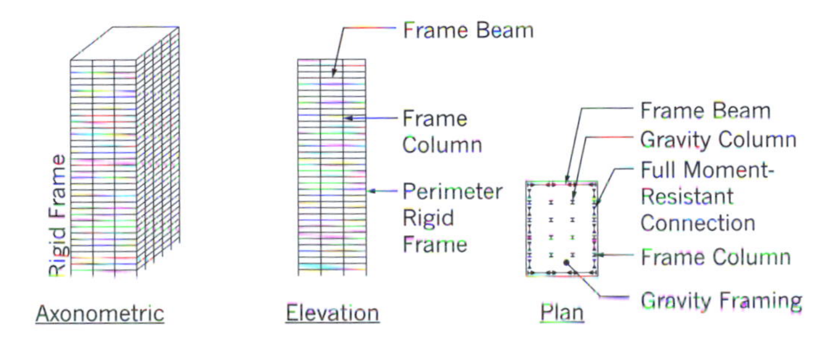

General limit of number of stories = 35 and height = 142 meters (465 ft, 9 in)

The moment-resisting frame typically comprises wide-flanged shaped columns and beams with primary frame connections typically bolted with slip-critical bolts, welded, or a combination of welded/bolted. Frame connections are fully fixed, developing the full bending moment capacity of the beams. Column spacing typically ranges from approximately the floor-to-floor height to twice the floor-to-floor height; generally, spacing varies between 4.5 m (15 ft) and 9 m (30 ft). The taller the building, the greater the need to use deeper wide-flanged sections for beams and columns and closer column spacing. Columns typically vary from W14 to W36 sections and beams typically vary from W21 to W36 sections. This moment-resisting frame has the same characteristics of the steel-rigid frame except frame connections are fully fixed, developing the full bending moment capacity of the beams with connections typically bolted with slip-critical bolts, welded, or a combination of welded/bolted.

8.2.1.3 Steel Frame with Shear Truss

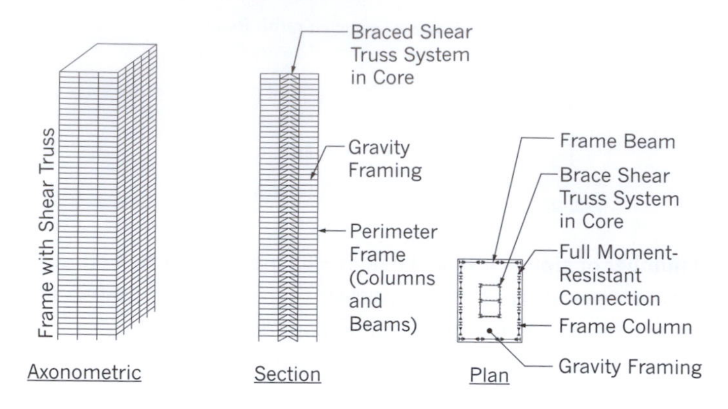

General limit of number of stories = 50 and height = 202 meters (662 ft, 6 in)

The shear truss, typically consisting of diagonal wide-flanged, angle, T-shaped, or tubular members, is located in the central core area or at the perimeter. Chevron (or k-braced) or concentrically braced (x-braced) frames are typically used. Gusset plates are typically used to connect intersecting members. Columns are typically spaced 4.5 m (15 ft) to 9 m (30 ft) on-center.

The frame is typically comprised of wide-flanged shaped columns and beams with primary frame connections typically bolted, welded, or a combination of welded/bolted. Frame connections are fully fixed, developing the full bending moment capacity of the beams. Column spacing typically ranges from approximately the floor-to-floor height to twice the floor-to-floor height; generally, spacing varies between 4.5 m (15 ft) and 9 m (30 ft). The taller the building, the greater the need to use deeper wide-flanged sections for beams and columns and closer column spacing. Columns typically vary from W14 to W36 sections and beams typically vary from W21 to W36 sections.

Shear truss is typically used to resist lateral loads for strength and drift control. Moment-resisting frames are used to supplement lateral load strength and drift control, providing increased torsional resistance of the structural system.

8.2.1.4 Steel Frame with Shear, Belt, and Outrigger Trusses

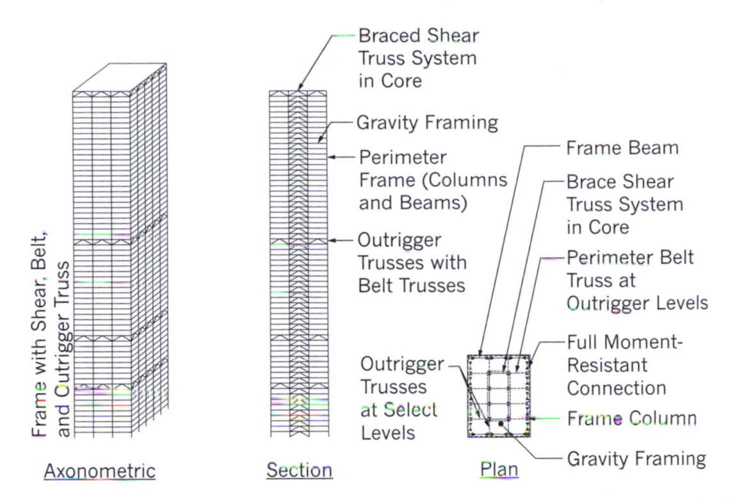

General limit of number of stories = 65 and height = 266 meters (872 ft, 6 in)

The shear truss, typically consisting of diagonal wide-flanged, angle, T-shaped, or tubular members, is located in the central core area or at the perimeter. Chevron (or k-braced) or concentrically braced (x-braced) frames are typically used. Gusset plates are typically used to connect intersecting members. Columns are typically spaced 4.5 m (15 ft) to 9 m (30 ft) on-center.

Outrigger trusses, typically at least one story tall (preferably two stories tall), are used to interconnect the central core and perimeter frame. Outrigger trusses typically consist of large wide-flanged or built-up members.

Belt trusses, located within the perimeter frame at the same level as the outrigger trusses, are used to distribute forces fairly evenly from the outrigger trusses to the perimeter frame.

The frame typically comprises wide-flanged shaped columns and beams with primary frame connections typically bolted, welded, or a combination of welded/bolted. Frame connections are fully fixed, developing the full bending moment capacity of the beams. Column spacing typically ranges from approximately the floor-to-floor height to twice the floor-to-floor height; generally, spacing varies between 4.5 m (15 ft) and 9 m (30 ft). The taller the building, the greater the need to use deeper wide-flanged sections for beams and columns and closer column spacing. Columns typically vary from W14 to W36 sections and beams typically vary from W21 to W36 sections.

Shear truss is typically used to resist lateral loads for strength and drift control. Moment-resisting frames are used to supplement lateral load strength and drift control providing increased torsional resistance of the structural system. Outrigger trusses, generally located at 25% of the height, 50% of the height, and top of the structure, act as levers, restricting the displacements of the core and introducing axial loads into columns of the perimeter frame.

8.2.1.5 Steel Exterior Framed Tube

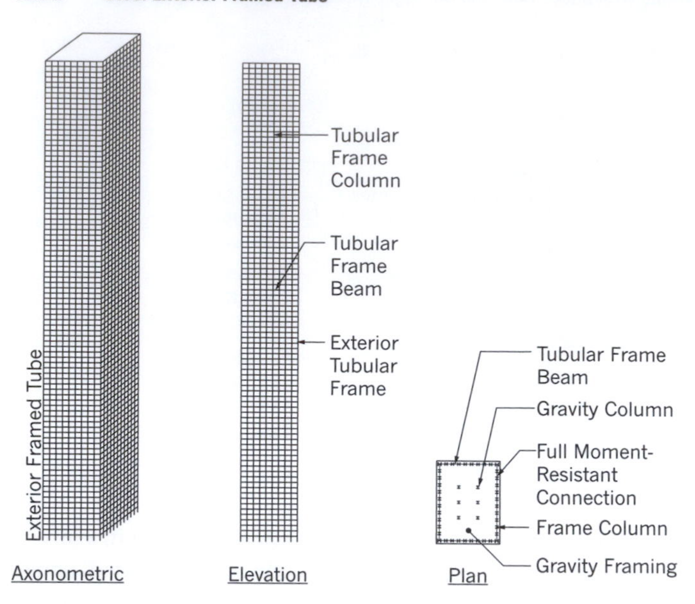

General limit of number of stories = 90 and height = 366 meters (1200 ft, 6 in)

The tubular frame typically comprises wide-flanged shaped or built-up columns and beams with primary frame connections typically bolted, welded, or a combination of welded/bolted. Frame connections are fully fixed, developing the full bending moment capacity of the beams. Column spacing approximately matches the floor-to-floor height, while aligning with interior partitions/exterior wall mullion modules (column spacing typically 4.5 m (15 ft)). The taller the building, the greater the need to use deeper wide-flanged sections for beams and columns and potentially closer column spacing.

The tubular frame is engineered to approximately equalize the bending stiffness of the columns and the beams. The structural system attempts to equalize axial load (reduce shear lag) attracted to columns when the overall structure is subjected to lateral loads. Shear lag is the phenomenon that describes the unequal distribution of force on the leading or back face columns when the frame is subjected to lateral loads.

8.2.1.6 Steel Bundled Frame Tube

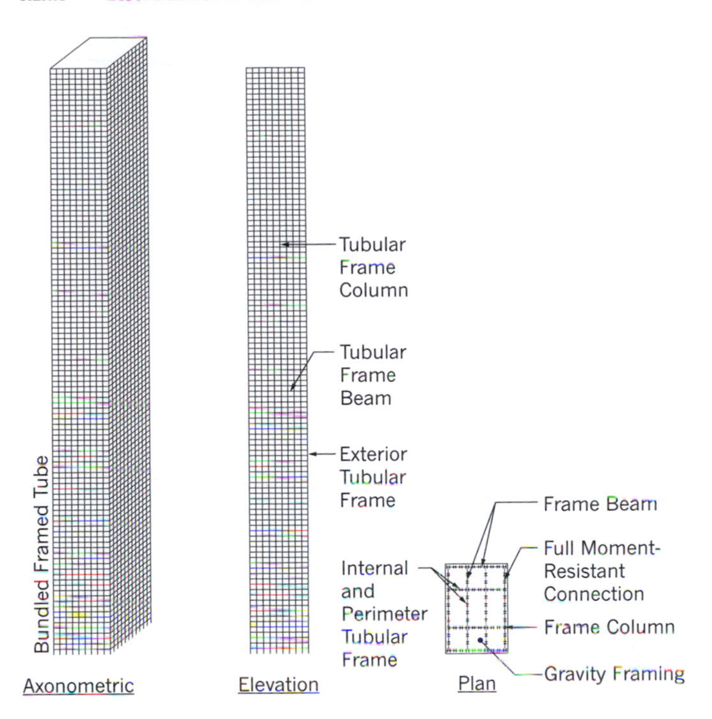

General limit of number of stories = 110 and height = 446 meters (1462 ft, 10 in)

The tubular frame typically comprises wide-flanged shaped or built-up columns and beams with primary frame connections typically bolted, welded, or a combination of welded/bolted. Frame connections are fully fixed developing the full bending moment capacity of the beams. Column spacing approximately matches the floor-to-floor height while aligning with interior partitions/exterior wall mullion modules (column spacing typically 4.5 m (15 ft)). The taller the building, the greater the need to use deeper wide-flanged sections for beams and columns and potentially closer column spacing.

The tubular frame is engineered to approximately equalize the bending stiffness of the columns and the beams. The structural system attempts to equalize axial load (reduce shear lag) attracted to columns when the overall structure is subjected to lateral loads. Shear lag is the phenomenon that describes the unequal distribution of force on the leading or back face columns when the frame is subjected to lateral loads.

The bundled tube uses a cellular concept, introducing interior frames, to further reduce shear lag. These bundle tubes may include belt trusses at levels where floor plans transition from large to small in order to interconnect or tie the tubular frames together.

8.2.1.7 Steel Exterior Diagonal Tube

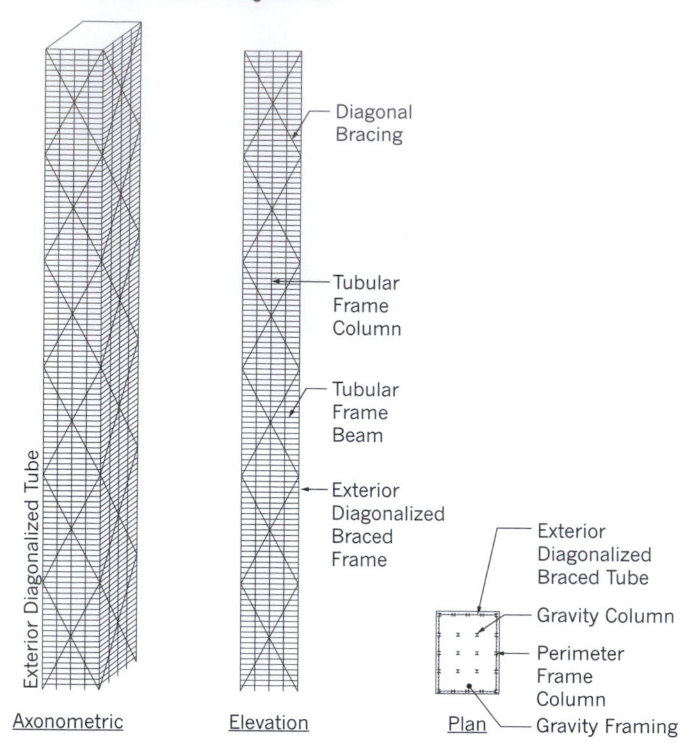

General limit of number of stories = 120 and height = 486 meters (1594 ft, 1 in)

The exterior diagonal tube typically comprises wide-flanged shaped or built-up columns and beams with primary frame connections typically bolted, welded, or a combination of welded/bolted. Frame connections are fully fixed developing the full bending moment capacity of the beams. Column spacing approximately matches the floor-to-floor height, while aligning with interior partitions/exterior wall mullion modules (column spacing typically 4.5 m (15 ft)). The taller the building, the greater the need to use deeper wide-flanged sections for beams and columns and potentially closer column spacing.

The exterior diagonal tube introduces diagonal members into the perimeter of the structure. These diagonal members typically exist on multiple-floor intervals. The structural system attempts to equalize axial load (reduce shear lag) attracted to columns when the overall structure is subjected to lateral loads. Shear lag is the phenomenon that describes the unequal distribution of force on the leading or back face columns when the frame is subjected to lateral loads. The use of diagonal members in the tube significantly increases structural efficiency (the use of materials) since the behavior is dominated by axial rather than bending behavior.

8.2.1.8 Steel Superframe

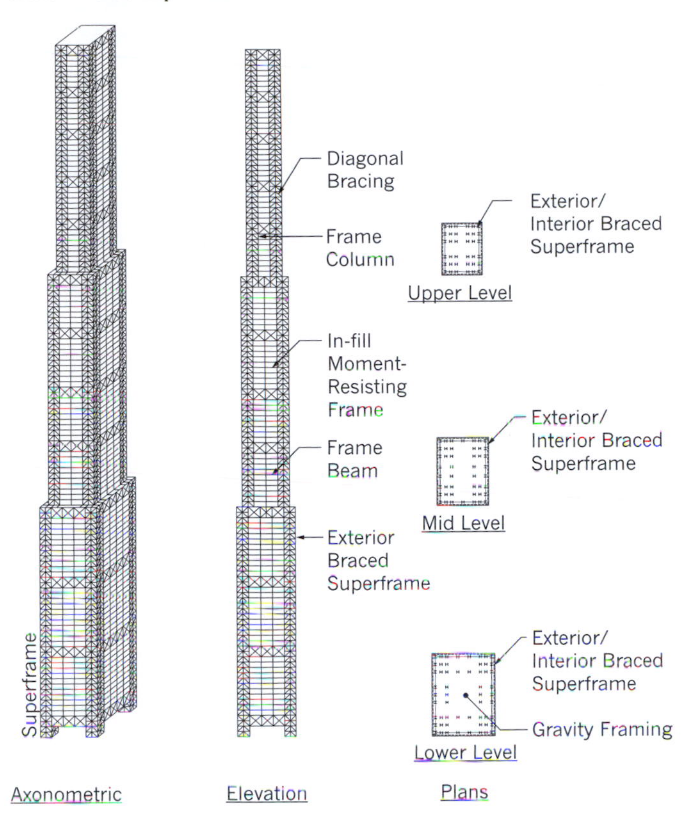

General limit of number of stories = 130 and height = 526 meters (1725 ft, 3 in)

The exterior diagonal frame is typically comprised of wide-flanged shaped or built-up columns and beams with primary frame connections typically bolted, welded, or a combination of welded/bolted. Frame connections are fully fixed, developing the full bending moment capacity of the beams. Column spacing approximately matches the floor-to-floor height while aligning with interior partitions/exterior wall mullion modules (column spacing typically 4.5 m (15 ft)). The taller the building, the greater the need to use deeper wide-flanged sections for beams and columns and potentially closer column spacing.

The perimeter frame is configured on a mega or superframe module with horizontal and vertical members of the frame existing in multiple bays and multiple stories. Each bay of the perimeter frame is diagonalized. The system allows for plan transitions along the height, and incorporates inherent belt trusses at transition floors. The concept allows for large atria or openings through the structure and allows for winds to pass through the structure to reduce the shear and overturning demand on the structure.

8.2.1.9 Steel Diagonal Mesh Tube Frame

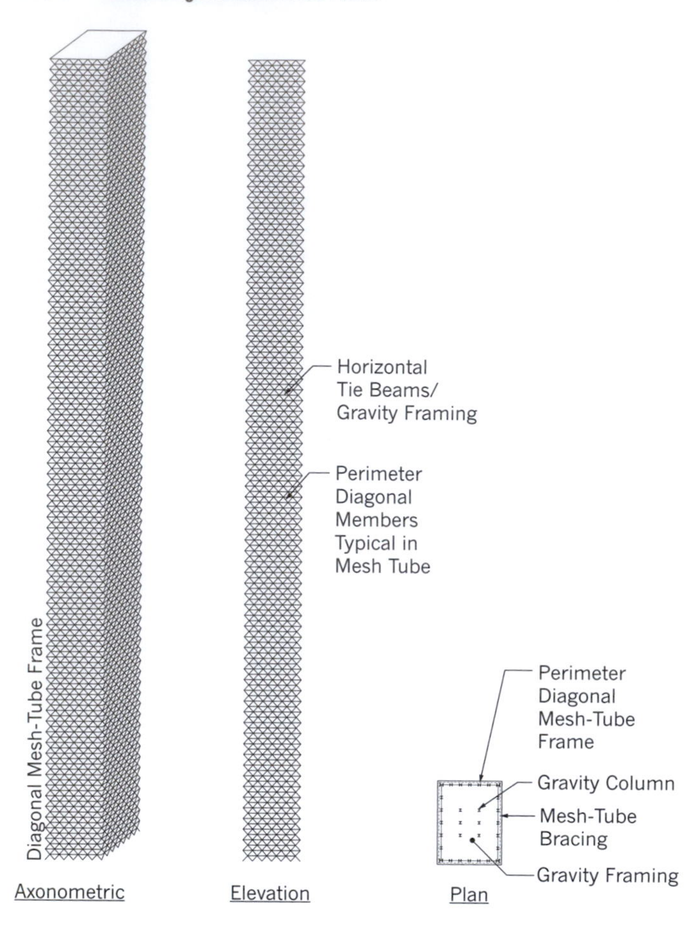

General limit of number of stories = 150 and height = 606 meters (1987 ft, 7 in)

The exterior diagonal mesh tube frame is typically comprised of wide-flanged shaped or built-up columns and beams with primary frame connections typically bolted, welded, or a combination of welded/bolted. Frame connections are fully fixed, developing the full bending moment capacity of the beams. Column spacing approximately matches the floor-to-floor height while aligning with interior partitions/exterior wall mullion modules (column spacing typically 4.5 m (15 ft)). The taller the building, the greater the need to use deeper wide-flanged sections for beams and columns and potentially closer column spacing.

The perimeter frame is configured to only introduce axial load into diagonal frame members. Bending moments on individual members are essentially eliminated, significantly increasing structural efficiency while minimizing material quantities. Tighter weave of the mesh results in smaller structural members and even greater efficiency.

8.2.2 Reinforced Concrete

Building heights listed as follows are based on an average floor-to-floor height of 3.2 m (10 ft, 6 in) allowing for a 2.45 m (8 ft) tall ceiling. Six meters (20 ft) are included for lobby spaces. One 6.4 meter (21 ft) tall space included in the mid-height of the tower for mechanical spaces or a sky lobby for buildings 60 stories and above.

8.2.2.1 Concrete Frame

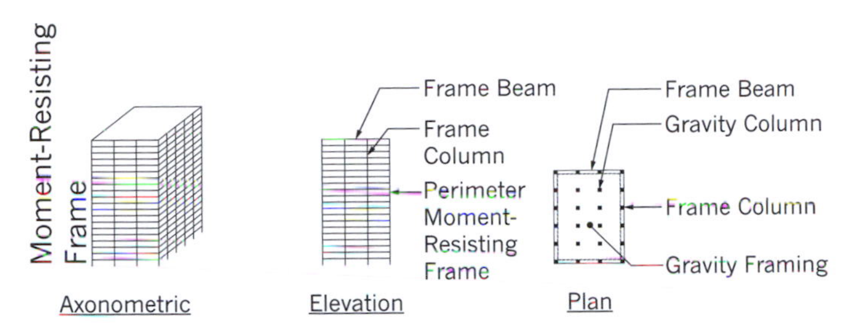

General limit of number of stories = 20 and height = 66.8 meters (219 ft, 1 in)

The moment-resisting frames are typically comprised of rectangular or square columns and rectangular beams. Frame joints use rebar detailing to develop the full bending moment capacity of the beams. Column spacing typically ranges from approximately the floor-to-floor height to twice the floor-to-floor height; generally, spacing varies between 4.5 m (15 ft) and 9 m (30 ft). The taller the building, the greater the need to use deeper rectangular sections for beams and columns and closer column spacing, with rectangular column sections orientated to provide greatest bending resistance.

8.2.2.2 Concrete Shear Wall

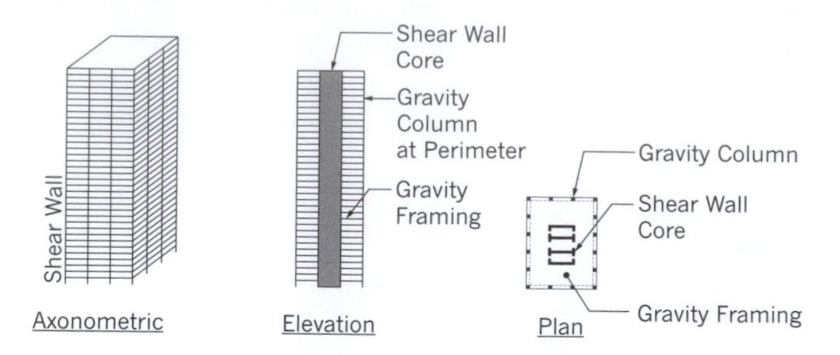

General limit of number of stories = 35 and height = 114.8 meters (376 ft, 6 in)

Shear walls within a tower are most commonly centrally located around service areas including elevators, mechanical spaces, and restrooms. However, these walls may be distributed throughout the floor plan in structures with residential programs. In some cases shear walls are located eccentrically in the floor plan and must resist significant torsion due to eccentrically applied wind loads or seismic loads due to eccentric relationship between center of mass and lateral stiffness.

Shear wall spacing locations vary, but are generally located 9 m (30 ft) apart to allow for a double bank of elevators and an elevator lobby. Link beams are used to interconnect wall segments where doors or mechanical openings are required in the core. The link beam depths are generally maximized to obtain greatest shear and bending resistance and must be coordinated with doorway heights and mechanical systems.

Shear walls resist all lateral loads and can be subjected to net tension and foundation uplift/overturning. Therefore, plan location/size and gravity load balancing is important to minimize any tension on shear wall elements.

8.2.2.3 Concrete Frame—Shear Wall

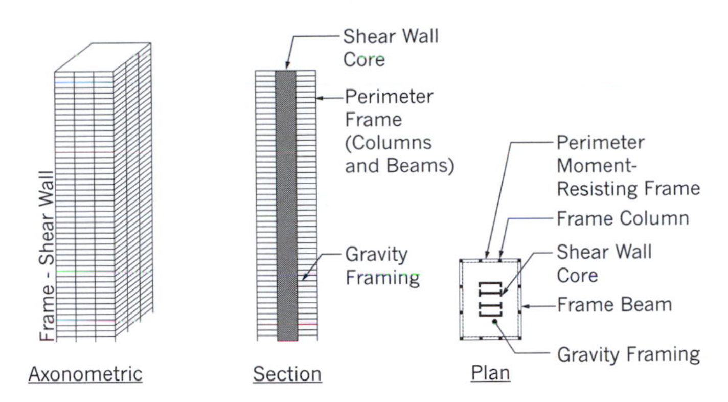

General limit of number of stories = 50 and height = 162.8 meters (534 ft)

Shear walls within a tower are most commonly centrally located around service areas, including elevators, mechanical spaces, and restrooms. However, these walls may be distributed throughout the floor plan in structures with residential programs. In some cases, shear walls are located eccentrically in the floor plan and must resist significant torsion due to eccentrically applied wind loads or seismic loads due to eccentric relationship between center of mass and lateral stiffness.

Shear wall spacing locations vary, but are generally located 9 m (30 ft) apart to allow for a double bank of elevators and an elevator lobby. Link beams are used to interconnect wall segments where doors or mechanical openings are required in the core. The link beam depths are generally maximized to obtain greatest shear and bending resistance and must be coordinated with doorway heights and mechanical systems.

Frames are combined with the shear walls to increase strength and stiffness. The moment-resisting frames are typically comprised of rectangular or square columns and rectangular beams. Frame joints use rebar detailing to develop the full bending moment capacity of the beams. Column spacing typically ranges from approximately the floor-to-floor height to twice the floor-to-floor height; generally, spacing varies between 4.5 m (15 ft) and 9 m (30 ft). The taller the building, the greater the need to use deeper rectangular sections for beams and columns and closer column spacing, with rectangular column sections orientated to provide greatest bending resistance.

Shear walls resist a majority of the lateral load, especially in lower portions of the structure, and can be subjected to net tension and foundation uplift/overturning. Therefore, plan location/size and gravity load balancing is important to minimize any tension on shear wall elements.

8.2.2.4 Concrete Framed Tube

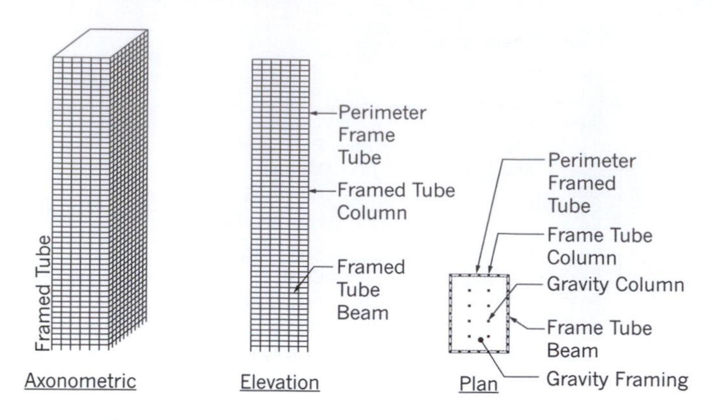

General limit of number of stories = 55 and height = 178.8 meters (586 ft, 6 in)

Moment-resisting tubular frames are typically comprised of rectangular or square columns and rectangular beams. Frame joints use rebar detailing to develop the full bending moment capacity of the beams. Column spacing typically ranges from a dimension slightly less than the floor-to-floor height to approximately equal to the floor-to-floor height; generally, spacing varies between 3.0 m (10 ft) and 4.5 m (15ft). The taller the building, the greater the need to use deeper rectangular sections for beams and columns and closer column spacing, with rectangular column sections orientated to provide greatest bending resistance.

The tubular frames are engineered to approximately equalize the bending stiffness of the columns and the beams. The structural system attempts to equalize axial load (reduce shear lag) attracted to columns when the overall structure is subjected to lateral loads. Shear lag is the phenomenon that describes the unequal distribution of force on the leading or back face columns when the frame is subjected to lateral loads.

8.2.2.5 Concrete Tube-in-Tube

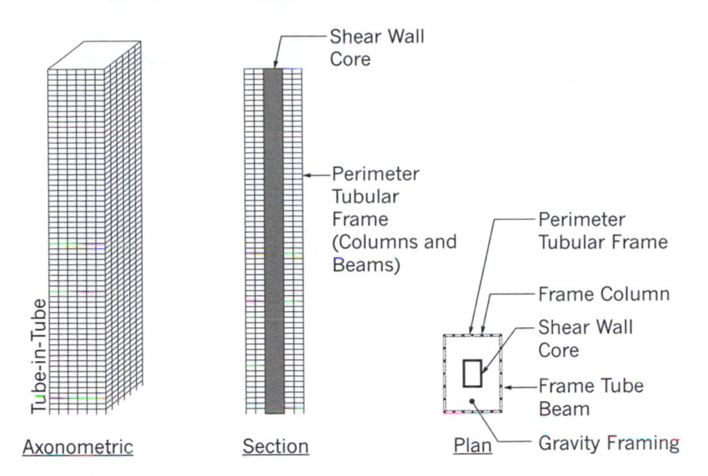

General limit of number of stories = 65 and height = 214 meters (701 ft, 11 in)

Moment-resisting tubular frames are typically comprised of rectangular or square columns and rectangular beams. Frame joints use rebar detailing to develop the full bending moment capacity of the beams. Column spacing typically ranges from a dimension slightly less than the floor-to-floor height to approximately equal to the floor-to-floor height; generally, spacing varies between 3.0 m (10 ft) and 4.5 m (15 ft). The taller the building, the greater the need to use deeper rectangular sections for beams and columns and closer column spacing, with rectangular column sections orientated to provide greatest bending resistance.

The tubular frames are engineered to approximately equalize the bending stiffness of the columns and the beams. The structural system attempts to equalize axial load (reduce shear lag) attracted to columns when the overall structure is subjected to lateral loads. Shear lag is the phenomenon that describes the unequal distribution of force on the leading or back face columns when the frame is subjected to lateral loads.

Interior tubular frames or shear walls are combined with perimeter tubular frames. These frames or shear walls provide additional strength and stiffness to the tubular system.

8.2.2.6 Concrete Modular Tube

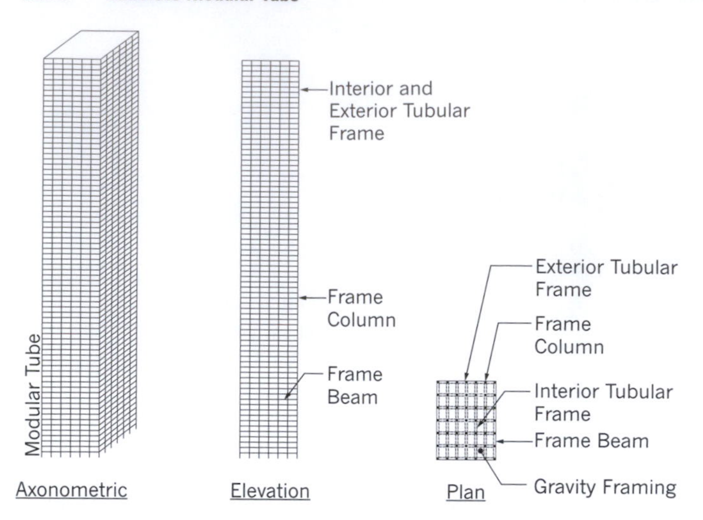

General limit of number of stories = 75 and height = 246 meters (806 ft, 10 in)

Moment-resisting tubular frames are typically comprised of rectangular or square columns and rectangular beams. Frame joints use rebar detailing to develop the full bending moment capacity of the beams. Column spacing typically ranges from a dimension slightly less than the floor-to-floor height to approximately equal to the floor-to-floor height; generally, spacing varies between 3.0 m (10 ft) and 4.5 m (15 ft). The taller the building, the greater the need to use deeper rectangular sections for beams and columns and closer column spacing, with rectangular column sections orientated to provide greatest bending resistance.

The tubular frames are engineered to approximately equalize the bending stiffness of the columns and the beams. The structural system attempts to equalize axial load (reduce shear lag) attracted to columns when the overall structure is subjected to lateral loads. Shear lag is the phenomenon that describes the unequal distribution of force on the leading or back face columns when the frame is subjected to lateral loads.

Interior tubular frames are combined with perimeter tubular frames. These frames are configured to develop a cellular configuration in plan. Interior frames must be carefully coordinated with occupied spaces since column spacing is limited on frame lines. These frames provide additional strength and stiffness to the tubular system.

8.2.2.7 Concrete Diagonal Braced Frame

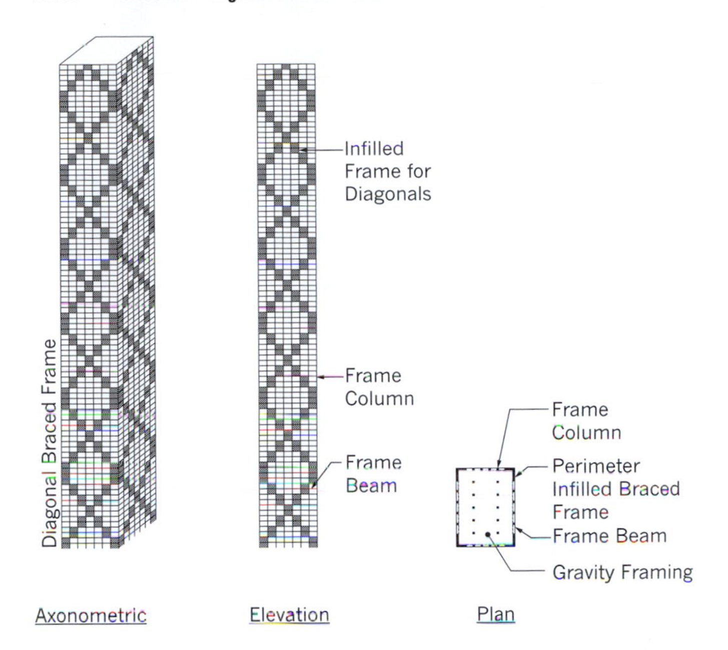

General limit of number of stories = 90 and height = 294 meters (964 ft, 4 in)

Moment-resisting frames are typically comprised of rectangular or square columns and rectangular beams. Frame joints use rebar detailing to develop the full bending moment capacity of the beams. Column spacing typically ranges from approximately the floor-to-floor height, to twice the floor-to-floor height; generally, spacing varies between 3.0 m (10 ft) and 4.5 m (15 ft). The taller the building, the greater the need to use deeper rectangular sections for beams and columns and closer column spacing, with rectangular column sections orientated to provide greatest bending resistance.

The tubular frames are infilled with a diagonal system at exterior locations. The structural system attempts to equalize axial load (reduce shear lag) attracted to columns when the overall structure is subjected to lateral loads. Shear lag is the phenomenon that describes the unequal distribution of force on the leading or back face columns when the frame is subjected to lateral loads.

8.2.2.8 Concrete Belt Shear Wall-Stayed Mast

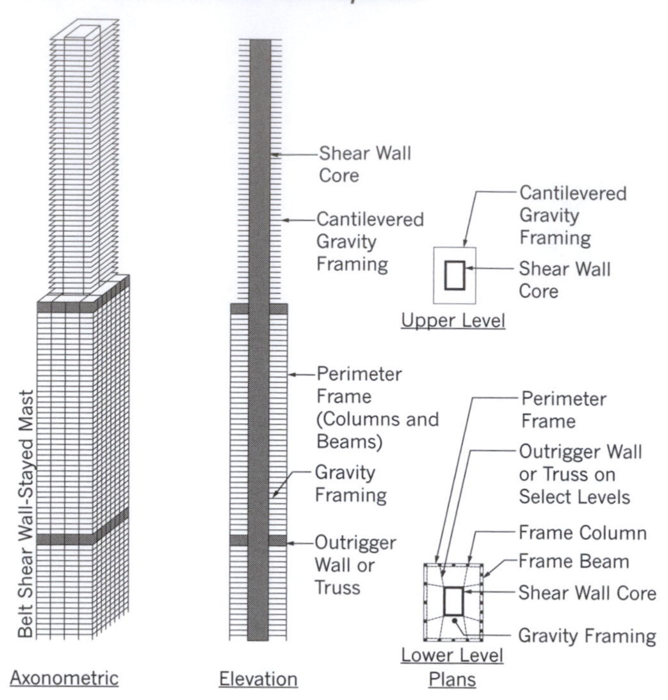

General limit of number of stories = 110 and height = 358 meters (1174 ft, 3 in)

Mega-core shear walls are centrally located in the floor plan around service areas including elevators, mechanical spaces, and restrooms. A closed form (square, rectangular, circular, octagon-shaped) is generally required to provide translational and torsional resistance.

Shear wall spacing locations vary, but are generally located 9 m (30 ft) apart to allow for a double bank of elevators and elevator lobby. Link beams are used to interconnect wall segments where doors or mechanical openings are required in the core. The link beam depths are generally maximized to obtain greatest shear and bending resistance and must be coordinated with doorway heights and mechanical systems.

Mega-core shear walls are interconnected to perimeter columns or frames with concrete walls (two stories tall typical) or in some cases steel trusses. These outrigger walls transfer loads from the central core to the perimeter columns or frames and maximize the structural depth of the floor plan.

The mega-core shear walls resist a majority of the lateral load, especially in lower portions of the structure, and can be subjected to net tension and foundation uplift/overturning. Therefore, plan location/size and gravity load balancing is important to minimize any tension on shear wall elements.

Belt walls may be used at outrigger levels to interconnect perimeter frame columns to maximize the axial stiffness of columns participating in the lateral system.

8.2.2.9 Concrete Diagonal Mesh Tube Frame

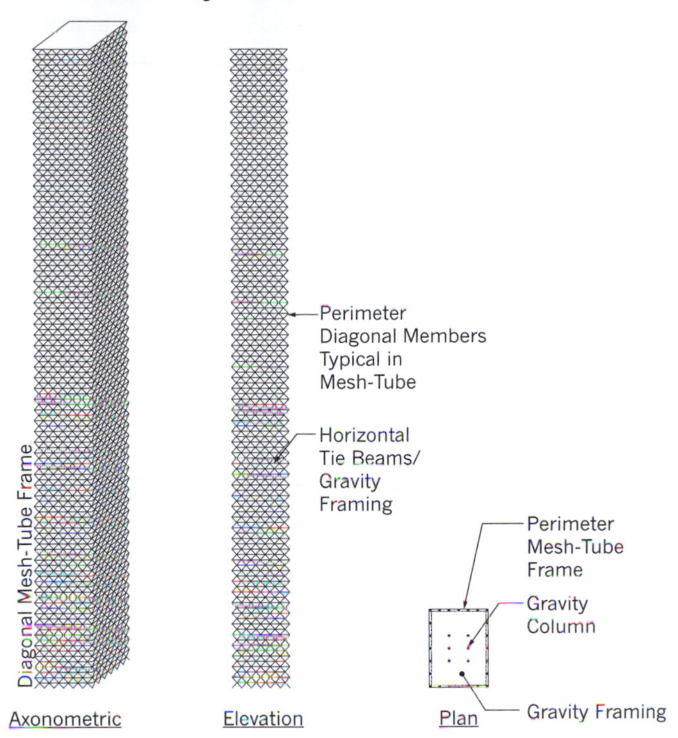

General limit of number of stories = 120 and height = 390 meters (1279 ft, 2 in)

The exterior diagonal mesh tube frame is typically comprised of rectangular or square diagonal members. Rectangular reinforced concrete frame beams are generally required at floor levels. Diagonal member spacing approximately matches the floor-to-floor height, while aligning with interior partitions/exterior wall mullion modules (diagonal spacing typically 4.5 m (15 ft)). The taller the building, the greater the need to use deeper diagonal sections and frame beams. A closer diagonal spacing is optional (3.0 m (10 ft)).

The perimeter frame is configured to introduce only axial load into diagonal frame members. Bending moments on individual members are essentially eliminated, significantly increasing structural efficiency while minimizing material quantities. Tighter weave of the mesh results in smaller structural members and even greater efficiency.

8.2.3 Composite (Combination of Steel and Concrete)

Building heights listed as follows are based on an average floor-to-floor height of 4 m (13 ft, 1½ in) allowing for a 2.75 m (9 ft) tall ceiling. Six meters (20 ft) are included for lobby spaces. One 8 meter (26 ft) tall space is included in the mid-height of the tower for mechanical spaces or a sky lobby for buildings 60 stories and above.

8.2.3.1 Composite Frame

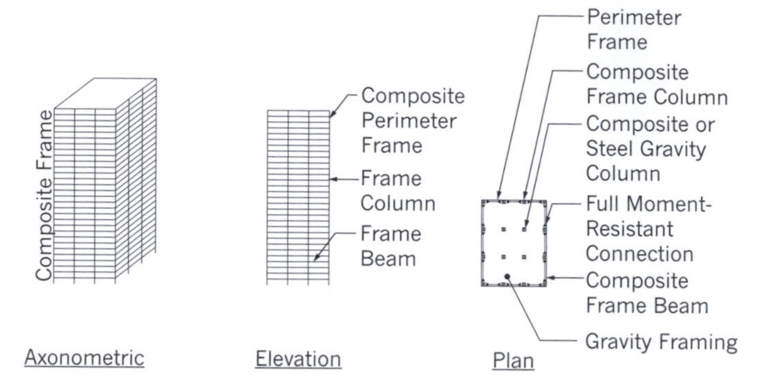

General limit of number of stories = 30 and height = 122 meters (400 ft, 2 in)

The moment-resisting frame is typically comprised of wide-flanged shaped columns and beams encased in reinforced concrete with primary frame connections typically bolted, welded, or a combination of welded/bolted and rebar detailed to develop full fixity and bending moment capacity of the beams. Columns and beams are typically encased in rectangular or square concrete sections. Column spacing typically ranges from approximately the floor-to-floor height to twice the floor-to-floor height (generally, spacing varies between 4.5 m (15 ft) and 9 m (30 ft)). The taller the building, the greater the need to use deeper wide-flanged sections for beams and columns and closer column spacing. Columns typically vary from W14 to W36 sections and beams typically vary from W21 to W36 sections. Structural steel within concrete sections varies from 3% to 10% of the concrete section.

8.2.3.2 Concrete Shear Wall—Steel Gravity Columns

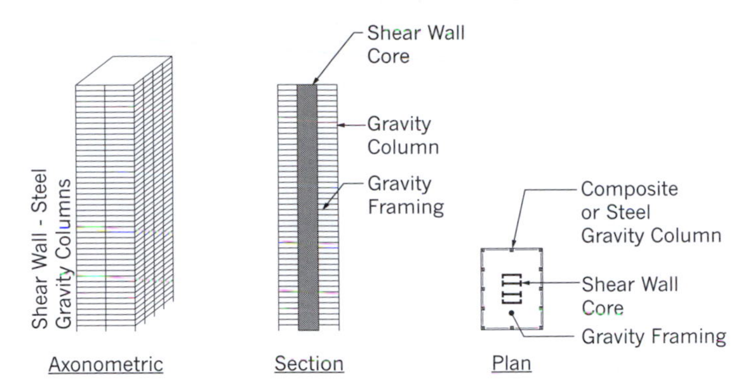

General limit of number of stories = 45 and height = 182 meters (597 ft)

Shear walls within a tower are most commonly centrally located around service areas including elevators, mechanical spaces, and restrooms. However, these walls may be distributed throughout the floor plan in structures with residential programs. In some cases shear walls are located eccentrically in the floor plan and must resist significant torsion due to eccentrically applied wind loads or seismic loads due to eccentric relationship between center of mass and lateral stiffness.

Shear walls resist all lateral loads and can be subjected to net tension and foundation uplift/overturning. Therefore, plan location/size and gravity load balancing is important to minimize any tension on shear wall elements. Steel columns resist gravity loads only from composite steel floor framing.

8.2.3.3 Concrete Shear Wall—Composite Frame

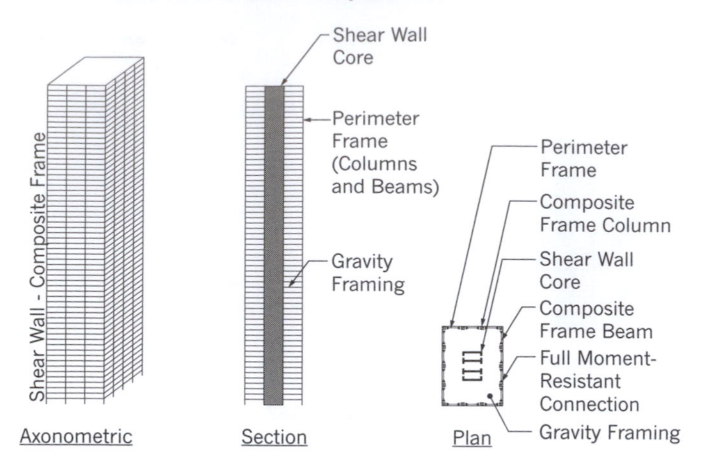

General limit of number of stories = 60 and height = 246 meters (806 ft, 10 in)

Shear walls within a tower are most commonly centrally located around service areas, including elevators, mechanical spaces, and restrooms. However, these walls may be distributed throughout the floor plan in structures with residential programs. In some cases, shear walls are located eccentrically in the floor plan and most resist significant torsion due to eccentrically applied wind loads or seismic loads due to eccentric relationship between center of mass and lateral stiffness.

Shear wall spacing locations vary, but are generally located 9 m (30 ft) apart to allow for a double bank of elevators and elevator lobby. Link beams are used to interconnect wall segments where doors or mechanical openings are required in the core. The link beam depths are generally maximized to obtain greatest shear and bending resistance and must be coordinated with doorway heights and mechanical systems.

Frames are combined with the shear walls to increase strength and stiffness. The moment-resisting frame is typically comprised of wide-flanged shaped columns and beams encased in reinforced concrete, with primary frame connections typically bolted, welded, or a combination of welded/bolted and rebar detailed to develop full fixity and bending moment capacity of the beams. Columns and beams are typically encased in rectangular or square concrete sections. Frame joints use rebar detailing to develop the full bending moment capacity of the beams. Column spacing typically ranges from approximately the floor-to-floor height to twice the floor-to-floor height; generally, spacing varies between 4.5 m (15 ft) and 9 m (30 ft). The taller the building, the greater the need to use deeper rectangular sections for beams and columns and closer column spacing, with rectangular column sections orientated to provide greatest bending resistance.

Shear walls resist a majority of the lateral load, especially in lower portions of the structure, and can be subjected to net tension and foundation uplift/overturning. Therefore, plan location/size and gravity load balancing is important to minimize any tension on shear wall elements.

8.2.3.4 Composite Tubular Frame

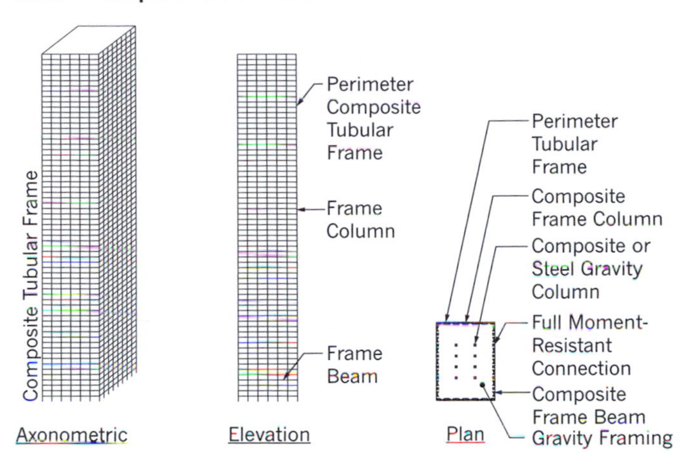

General limit of number of stories = 65 and height = 266 meters (872 ft, 6 in)

Moment-resisting tubular frames are typically comprised of structural steel wide-flange or built-up sections embedded in rectangular or square concrete columns and rectangular beams. Frame joints use bolted, welded, or a combination of bolted and welded steel connections and rebar detailing to develop the full bending moment capacity of the beams. Column spacing approximately matches the floor-to-floor height, while aligning with interior partitions/exterior wall mullion modules (column spacing typically 4.5 m (15 ft)). The taller the building, the greater the need to use deeper rectangular sections for beams and columns and closer column spacing, with rectangular column sections orientated to provide greatest bending resistance.

The tubular frames are engineered to approximately equalize the bending stiffness of the columns and the beams. The structural system attempts to equalize axial load (reduce shear lag) attracted to columns when the overall structure is subjected to lateral loads. Shear lag is the phenomenon that describes the unequal distribution of force on the leading or back face columns when the frame is subjected to lateral loads.

8.2.3.5 Composite Tube-in-Tube

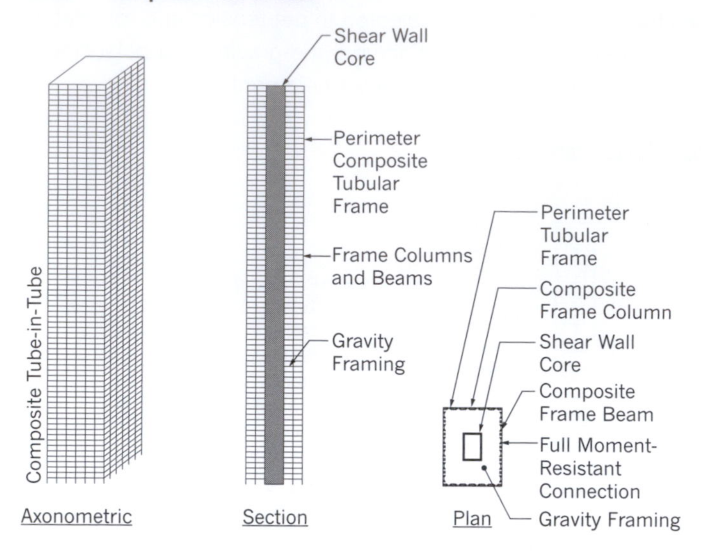

General limit of number of stories = 75 and height = 306 meters (1003 ft, 7 in)

Moment-resisting tubular frames are typically comprised of structural steel wide-flange or built-up sections embedded in rectangular or square concrete columns and rectangular beams. Frame joints use bolted, welded, or a combination of bolted and welded steel connections and rebar detailing to develop the full bending moment capacity of the beams. Column spacing approximately matches the floor-to-floor height, while aligning with interior partitions/exterior wall mullion modules (column spacing typically 4.5 m (15 ft)). The taller the building, the greater the need to use deeper rectangular sections for beams and columns and closer column spacing, with rectangular column sections orientated to provide greatest bending resistance.

The tubular frames are engineered to approximately equalize the bending stiffness of the columns and the beams. The structural system attempts to equalize axial load (reduce shear lag) attracted to columns when the overall structure is subjected to lateral loads. Shear lag is the phenomenon that describes the unequal distribution of force on the leading or back face columns when the frame is subjected to lateral loads.

Interior tubular frames or shear walls are combined with perimeter tubular frames. These frames or shear walls provide additional strength and stiffness to the tubular system.

8.2.3.6 Composite Modular Tube

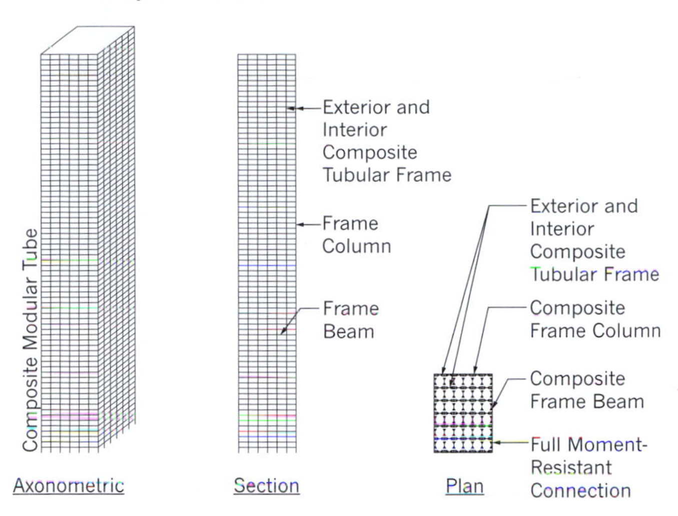

General limit of number of stories = 75 and height = 306 meters (1003 ft, 7 in)

Moment-resisting tubular frames are typically comprised of structural steel wide-flange or built-up sections embedded in rectangular or square concrete columns and rectangular beams. Frame joints use bolted, welded, or a combination of bolted and welded steel connections and rebar detailing to develop the full bending moment capacity of the beams. Column spacing approximately matches the floor-to-floor height, while aligning with interior partitions/exterior wall mullion modules (column spacing typically 4.5 m (15 ft)). The taller the building, the greater the need to use deeper rectangular sections for beams and columns and closer column spacing, with rectangular column sections orientated to provide greatest bending resistance.

The tubular frames are engineered to approximately equalize the bending stiffness of the columns and the beams. The structural system attempts to equalize axial load (reduce shear lag) attracted to columns when the overall structure is subjected to lateral loads. Shear lag is the phenomenon that describes the unequal distribution of force on the leading or back face columns when the frame is subjected to lateral loads.

Interior tubular frames are combined with perimeter tubular frames. These frames are configured to develop a cellular configuration in plan. Interior frames must be carefully coordinated with occupied spaces since column spacing is limited on frame lines. These frames provide additional strength and stiffness to the tubular system.

8.2.3.7 Composite Diagonal Braced Tube

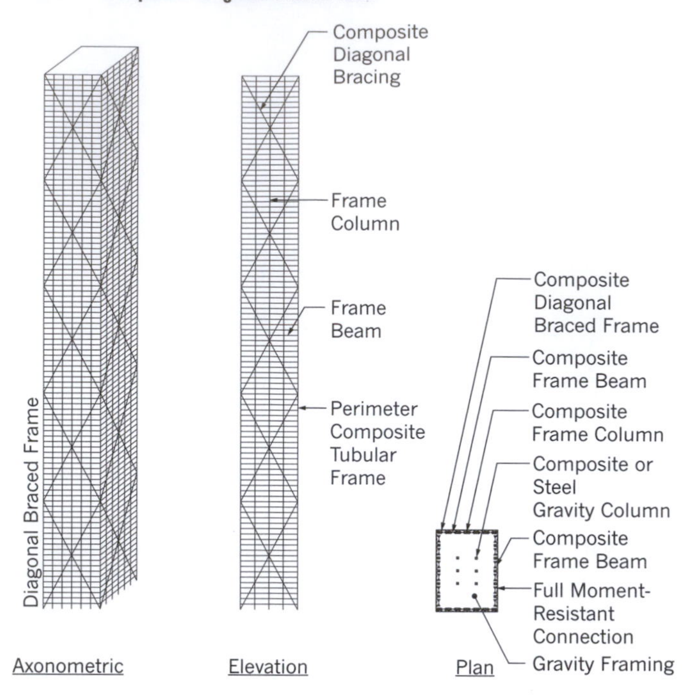

General limit of number of stories = 90 and height = 366 meters (1200 ft, 6 in)

The exterior diagonal tube is typically comprised of structural steel wide-flange or built-up sections embedded in rectangular or square concrete columns and rectangular beams. Frame joints use bolted, welded, or a combination of bolted and welded steel connections and rebar detailing to develop the full bending moment capacity of the beams. Column spacing approximately matches the floor-to-floor height, while aligning with interior partitions/exterior wall mullion modules (column spacing typically 4.5 m (15 ft)). The taller the building, the greater the need to use deeper wide-flanged sections for beams and columns and potentially closer column spacing.

The exterior diagonal tube introduces diagonal members into the perimeter of the structure. These diagonal members typically exist on multiple-floor intervals. The structural system attempts to equalize axial load (reduce shear lag) attracted to columns when the overall structure is subjected to lateral loads. Shear lag is the phenomenon that describes the unequal distribution of force on the leading or back face columns when the frame is subjected to lateral loads. The use of diagonal members in the tube significantly increases structural efficiency (the use of materials) since the behavior is dominated by axial rather than bending behavior.

8.2.3.8 Composite Belt Outrigger-Stayed Mast

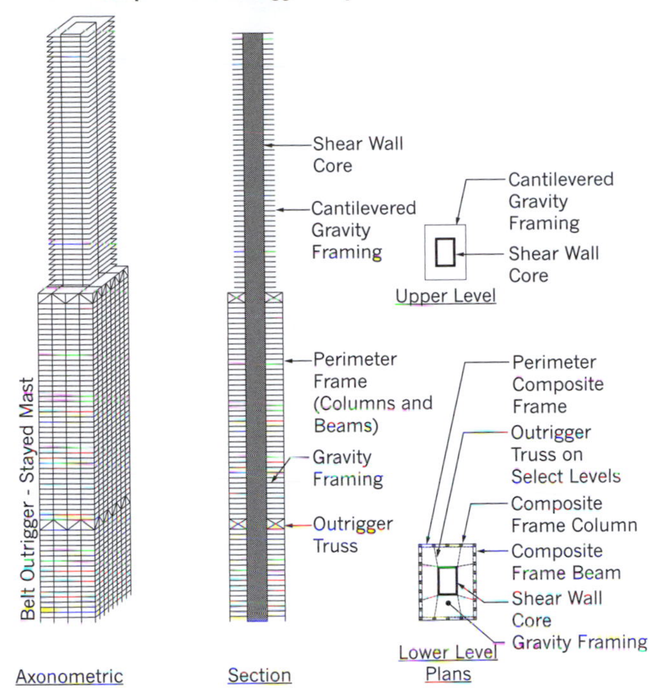

General limit of number of stories = 110 and height = 446 meters (1462 ft, 10 in)

Mega-core shear walls are located centrally in the floor plan located around service areas including elevators, mechanical spaces, and restrooms. A closed form (square, rectangular, circular, octagon-shaped) is generally required to provide translational and torsional resistance.

Shear wall spacing locations vary, but are generally located 9 m (30 ft) apart to allow for a double bank of elevators and elevator lobby. Link beams are used to interconnect wall segments where doors or mechanical openings are required in the core. The link beam depths are generally maximized to obtain greatest shear and bending resistance and must be coordinated with doorway heights and mechanical systems.

Mega-core shear walls are interconnected to perimeter columns or frames with steel trusses (two stories tall typical). These outrigger trusses transfer loads from the central core to the perimeter composite columns or frames and maximize the structural depth of the floor plan.

The mega-core shear walls resist a majority of the lateral load, especially in lower portions of the structure, and can be subjected to net tension and foundation uplift/overturning. Therefore, plan location/size and gravity load balancing is important to minimize any tension on shear wall elements.

Belt trusses may be used at outrigger levels to interconnect perimeter frame columns to maximize the axial stiffness of columns participating in the lateral system.

8.2.3.9 Composite Diagonal Mesh Tube Frame

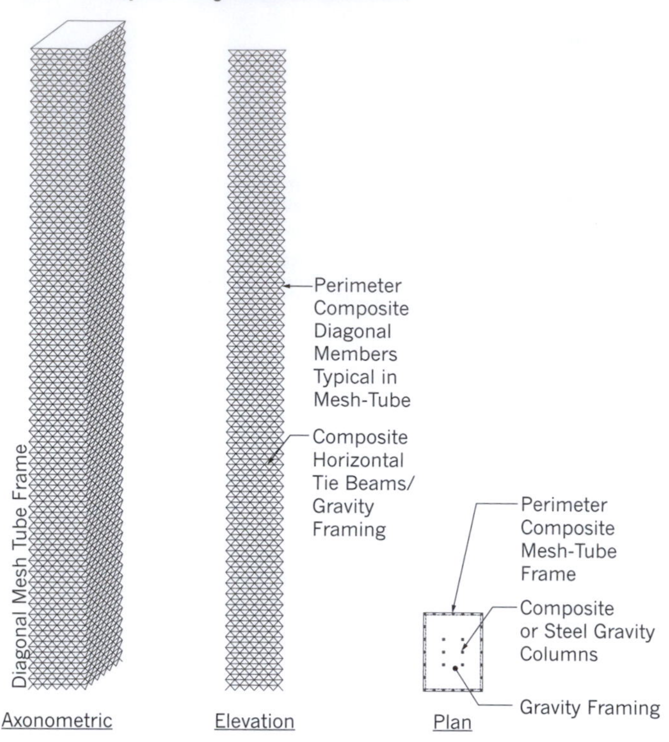

General limit of number of stories = 120 and height = 486 meters (1594 ft, 1 in)

The exterior diagonal mesh tube is typically comprised of structural steel wide-flange or built-up sections embedded in rectangular or square diagonal concrete members. Rectangular composite frame beams are generally required at floor levels. Diagonal member spacing approximately matches the floor-to-floor height, while aligning with interior partitions/ exterior wall mullion modules (diagonal spacing typically 4.5 m (15 ft)). The taller the building, the greater the need to use deeper diagonal sections and frame beams. A closer diagonal spacing is optional (3.0 m (10 ft)).

The perimeter frame is configured to introduce only axial load into frame members. Bending moments on individual members are essentially eliminated, significantly increasing structural efficiency while minimizing material quantities. Tighter weave of the mesh results in smaller structural members and even greater efficiency.

8.3 MAJOR SYSTEM DETAILS

Beyond typical details used for a majority of structural systems and components, unique systems details not only solve key structural engineering problems, but potentially become defining elements of the architecture.

8.3.1 The Arch—Broadgate Phase II

The articulated structural steel joint at the base of the tied arch in Broadgate Phase 11 (Exchange House), London located at each building face as well as two internal locations resolve compression (arch) and tension (tie) forces. The connection results in a single, and essentially pure, reaction applying only vertical loads to reinforced concrete end piers and foundations below. The joints at the exterior faces are offset from the occupied spaces/exterior walls and are fire-engineered to prove that fire-proofing is not required and allow for exposed, painted structural steel.

Construction and Final Photography, Broadgate Phase II (Exchange House), London, United Kingdom

8.3.2 The Rocker—Poly Corporation Headquarters, Beijing, China

A rocker or "reverse pulley" exists at Level 11 of the 22-story China Poly Headquarters in Beijing. The rocker allows for vertical hanging support of a museum through high-strength, spiral-wound bridge cables while allowing the building to move freely during a seismic event. Without the rocker system, the large cables would act as tension braces, developing forces from large displacements of the top of the building relative to the top of the museum. If the cables acted as braces, the forces developed could not be accommodated by the cables or the base building structure. In addition, the rocker and cable system provide lateral support for the world's largest cable net 60 m x 90 m (197 ft x 295 ft).

The Rocker, Poly Corporation Headquarters, Beijing, China

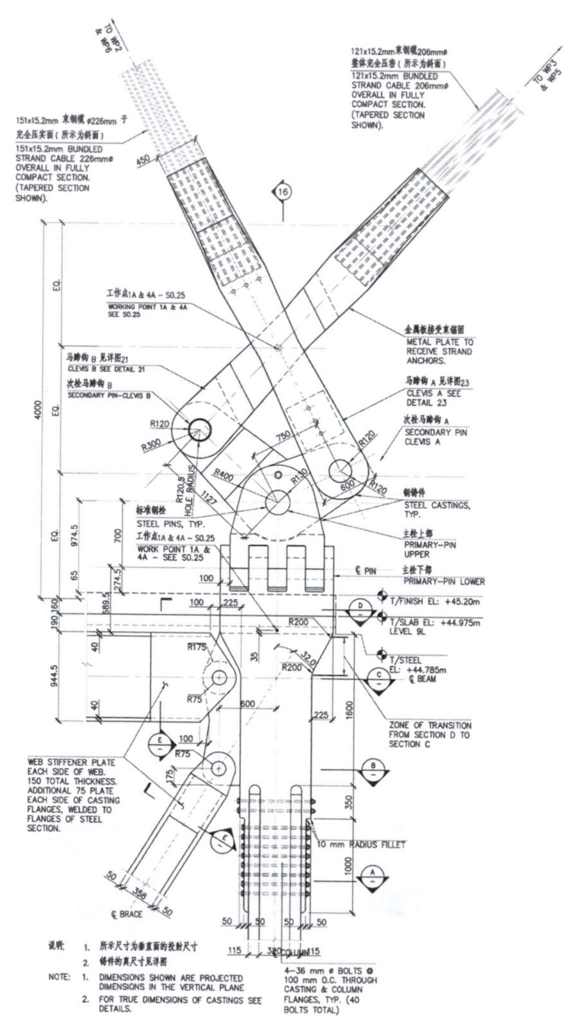

8.3.3 Pipe Collector—Kingtown International Center

A reinforced concrete frame is "wrapped" with a conventional steel pipe system, allowing for a 40% reduction in material that would otherwise be required for the lateral load resisting system for the Kingtown International Center, Nanjing, China. The pipes are eccentrically located at the corners of the structure, connected to steel elements encased in reinforced concrete columns. Since the diagonal pipe system only resists lateral load and the building is stable, even if part of the pipe system is lost to fire or another event, the pipes do not require fireproofing and can be simply painted and exposed (located in this structure between the double glass wall).

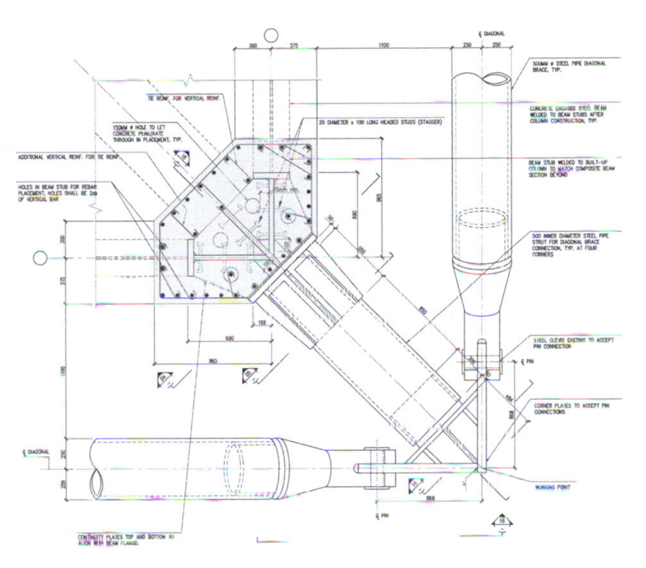

Pipe Collector, Bracing and Exterior Wall Construction, Kingtown International Center, Nanjing, China

8.3.4 Pinned Trusses—Jin Mao Tower

Pins installed within massive steel outrigger trusses in the Jin Mao Tower, Shanghai, allowed for free movement during construction when creep, shrinkage and elastic shortening imposed differential settlement between the central reinforced concrete core and perimeter composite mega-columns. Initially conceived from a simple model of popsicle sticks, tongue depressors, and wood dowels, the trusses act as free moving mechanisms through the time of relative displacements until high-strength bolts are installed and the structure is placed in full service, resisting all lateral loads.

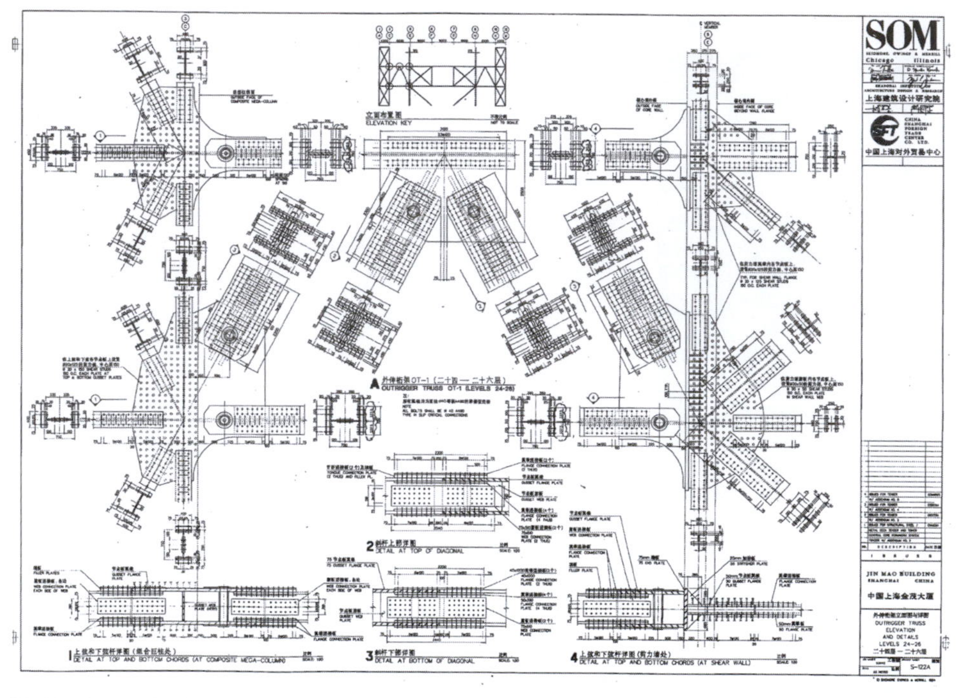

Working Drawing for Outrigger Truss, Construction Fit-Up at Shop, Lower Truss Connection Joint, Jin Mao Tower, Shanghai, China

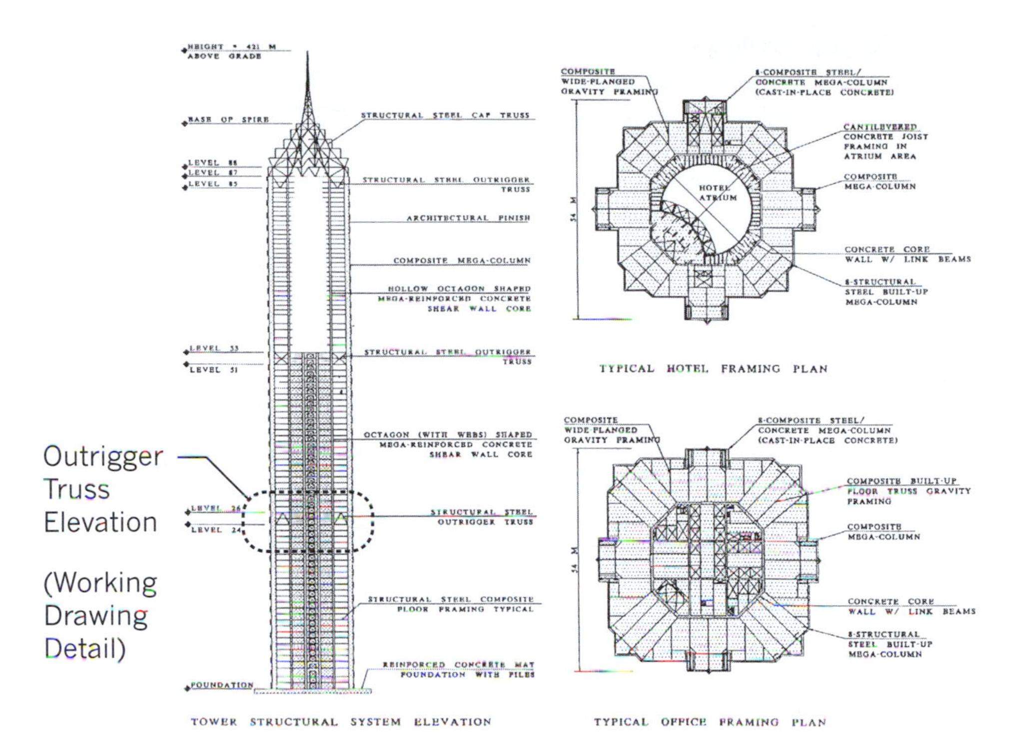

Structural System Elevation, Framing Plans, Overall Outrigger Truss Elevation, Detail of Exposed Panel

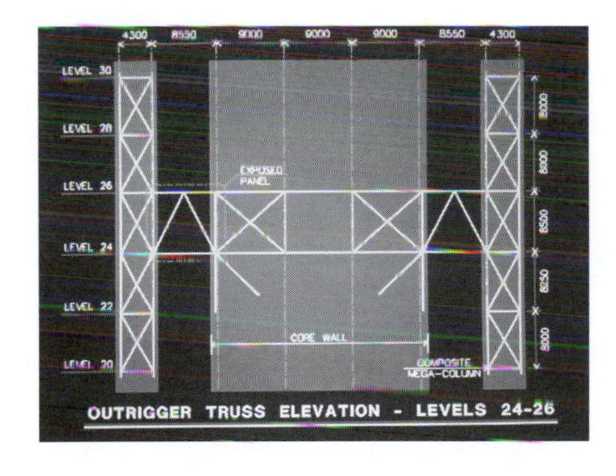

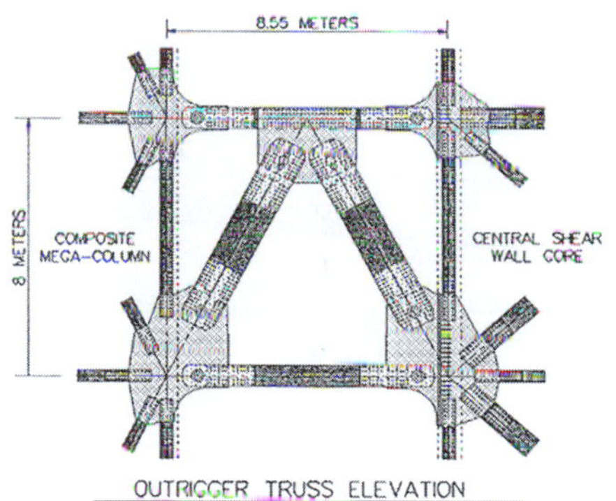

8.3.5 Diagonal Infill Panels

Before his untimely death in 1982, Fazlur Khan collaborated with Bruce Graham on perhaps his most important work, the 58-story Onterie Center Tower in Chicago, Illinois. Completed in 1986, the design incorporated many innovative ideas, most importantly using concepts of the braced frame originally conceived in steel by applying them to an alternative material, reinforced concrete.

Although Khan had previously used reinforced concrete for many of his innovative structures, (i.e. tubular frames) this was the first time a diagonal braced frame was created in concrete. The concepts of a reinforced concrete shear wall system that previously had been included in centrally located cores but now was used for infill panels to increase the stiffness of the frame.

These diagonally reinforced panels not only provided increased stiffness but simplified exterior window systems since the enclosure was only required in the regular frame units. The diagonal panels acted to evenly distribute gravity loads to counteract any potential tension due to lateral loads, reduce shear lag, and to create a clear load path to "channel-shaped frames" that exist at each end of the tower structure.

With all of the lateral loads resisted by the perimeter frame, no internal frames or cores were required, allowing for increased flexibility of occupied floor space used for apartments. Interior columns only needed to resist gravity loads so they could be placed where locations worked best with partitions, corridors, and service areas.

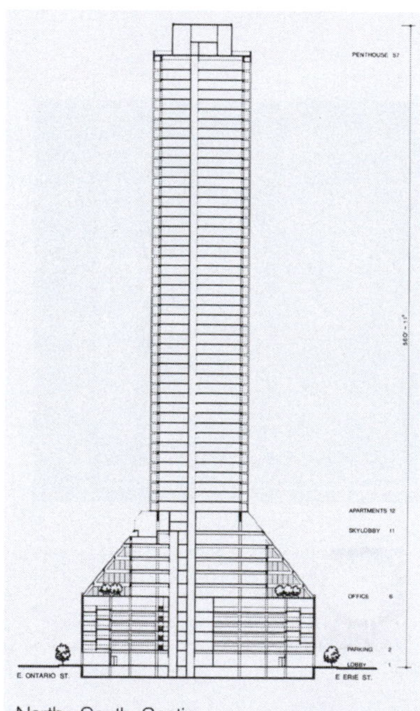

Ontarie Center, Chicago, IL

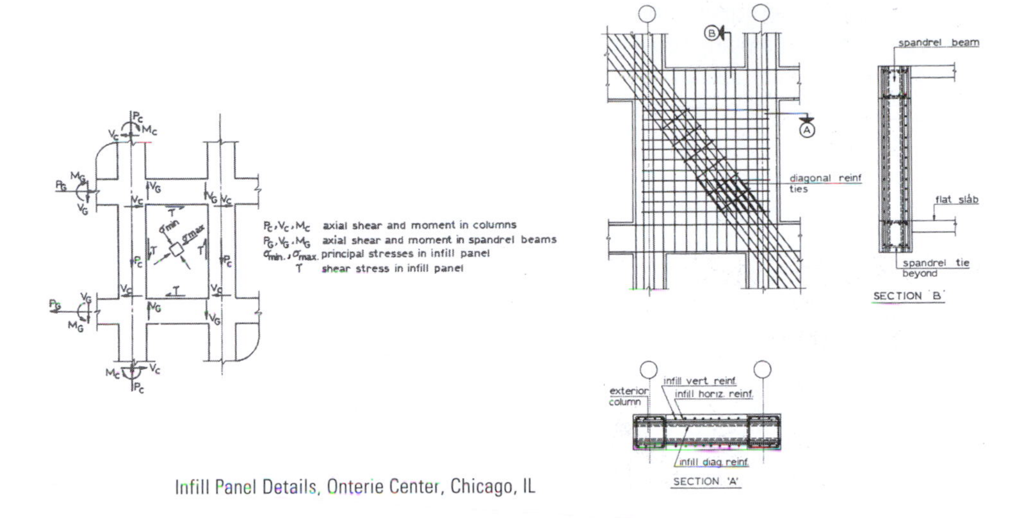

Infill Panel Details, Onterie Center, Chicago, IL

8.3.6 Thin Steel Plates—Tianjin Global Financial Center

In early concept design, the development group for the Tianjin Global Financial Center, Tianjin, China stated that they would like to use structural steel for the tower's structure. The goal of the design team was to use locally supplied material that was readily available. In the region, the manufacturing of thin steel plate for automobiles, military equipment, and shipbuilding was prevalent. The team imagined the design and construction of a 339 m (1112 ft) tall tower

Tallest Steel-plated Shear Wall Building in the World, Tianjin Global Financial Center, Tianjin, China

entirely out of thin steel plate. The central shear wall core, the circular columns, the built-up floor framing would all be designed and constructed of plate typically having a thickness of no more than 19 mm (¾ in).

The shear wall was designed considering tension field action incorporating vertical stiffeners to prevent plate buckling with columns and horizontal beams acting as boundary elements. The columns included cold-bent and vertical seem-welded sections and filled with concrete to increase axial capacity. Floor framing members where built up out of three plates and welded to form I-shaped framing members. The tower is the tallest steel-plated core wall building in the world.

Thin Steel Plates Used for Central Shear Wall, Columns, and Framing, Tianjin Global Financial Center, Tianjin, China

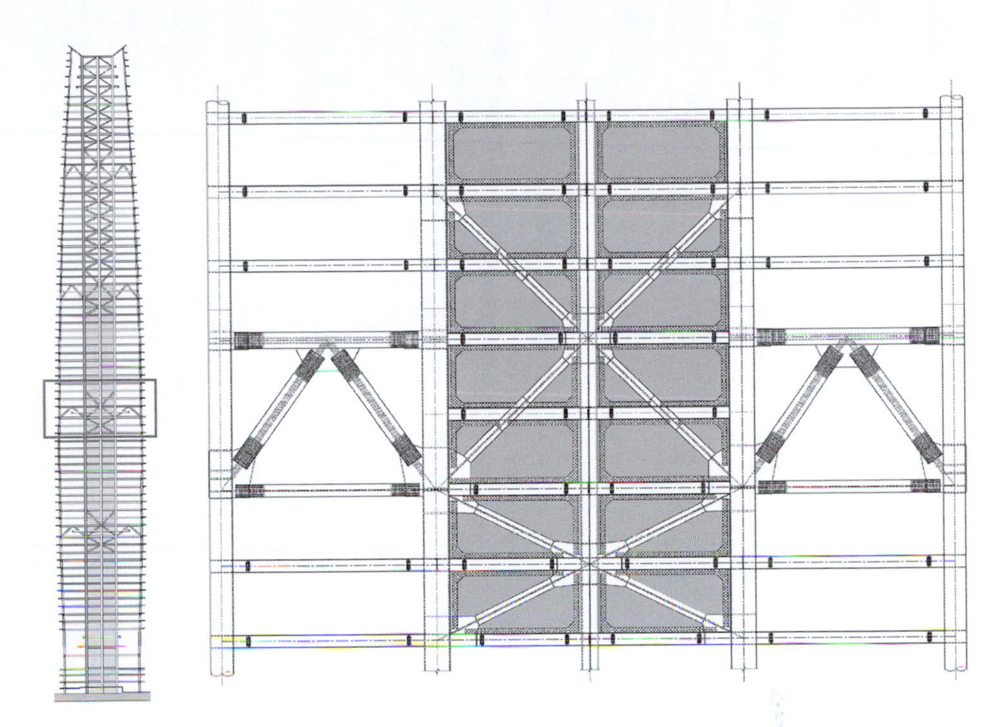

Steel-Plated Shear Wall Interconnected with Outrigger System, Tianjin Global Financial Center, Tianjin, China

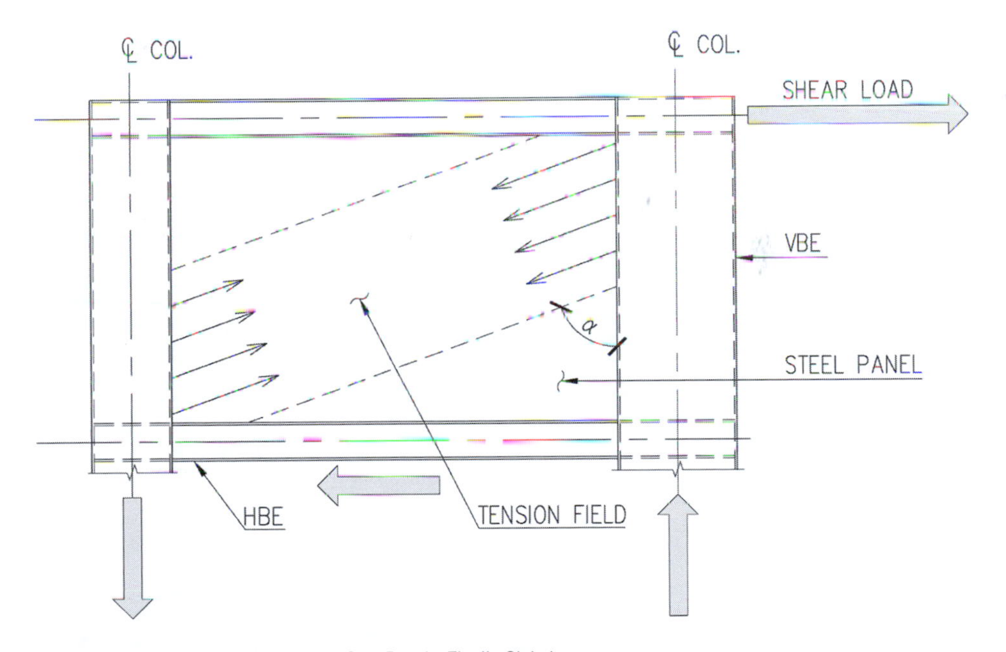

Tension Field Action Within Shear Wall Core Panels, Tianjin Global Financial Center, Tianjin, China

CHAPTER 9

NATURE

IN 1956, FRANK LLOYD WRIGHT conceived of the Mile-High Building. Since Wright, others have followed with concepts that approach one mile in height. Foster and Partners have proposed the Millennium Tower and Cervera & Pioz have proposed the Bionic Tower. Realistic steps toward this extreme height are incremental, with what was formerly the tallest building in the world, Taipei 101, designed by C.Y. Lee and Partners, at 508 m (1667 ft). At 828 m (2716 ft), the Burj Khalifa, designed by SOM and completed in 2010, has significantly surpassed this height.

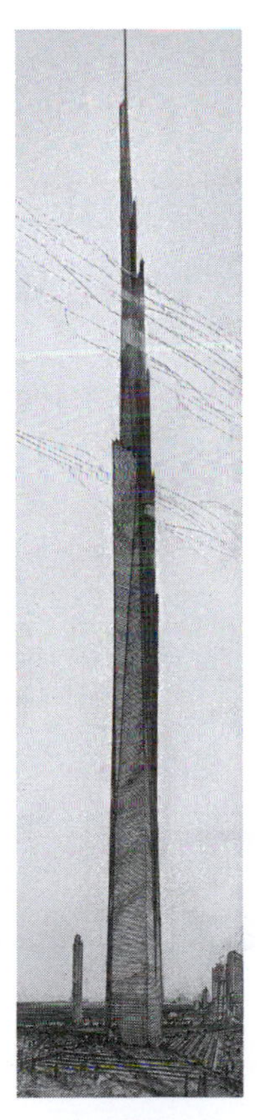

Tree Root Structure

The Illinois: The Mile High Building, Frank Lloyd Wright

FACING PAGE
Bamboo in Nature

Millennium Tower, Foster + Partners

Bionic Tower, Maria Rosa Cervera, Javier Gómez Pioz

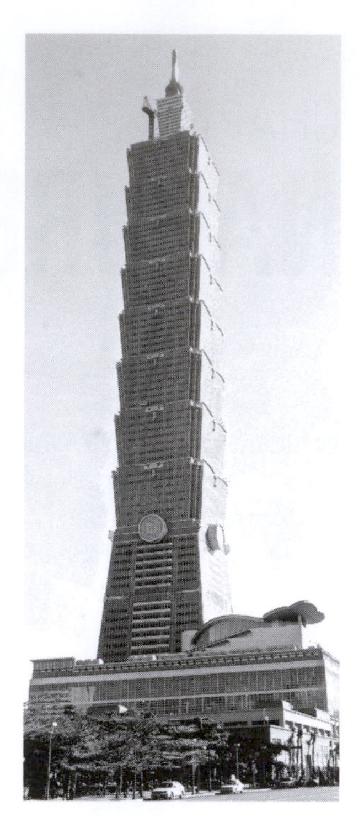

Taipei 101, C.Y. Lee & Partners

Variations on current structural systems and the development of new systems are important for the next generations of ultra-tall buildings. Variations developed to solve structural challenges while achieving extreme height often lead to interesting architectural solutions as well.

9.1 SCREEN FRAMES

Rigid frames for buildings can be stiffened to create further resistance to wind and seismic loads. The screen frames used on the Gemdale Plaza, Beijing, use lateral stiffening frames within a mega-frame concept. Economical moment-resisting frame heights can be increased from 35 stories to 50 stories.

The Rigid Frame,
Inland Steel Building, Chicago, IL

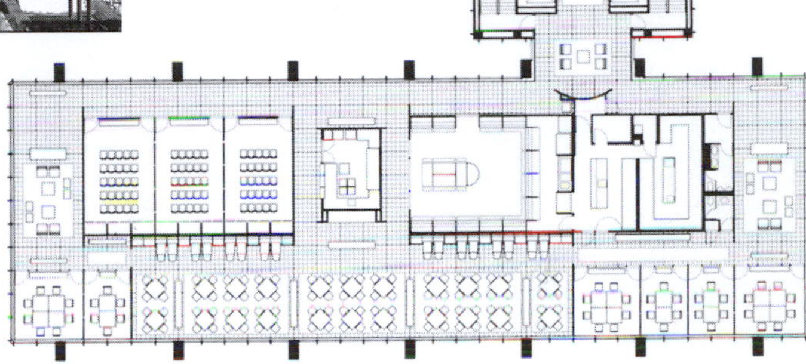

The Stiffened Screen Frame,
Gemdale Plaza, Beijing, China

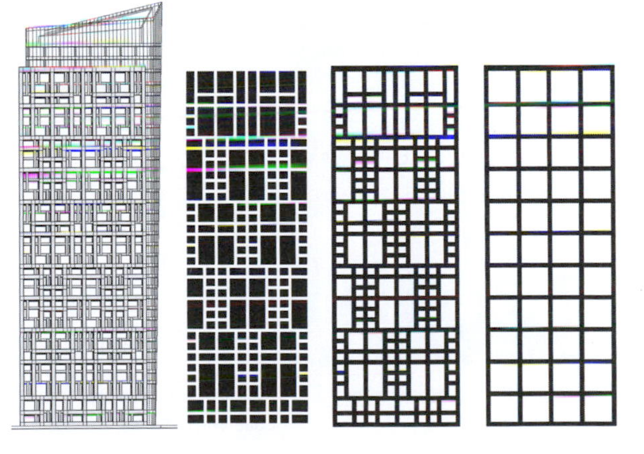

9.2 CORES AND PRESTRESSED FRAMES

The central core wall used to surround back-of-house areas, elevators, restrooms, and mechanical rooms is an excellent solution to tall building structures. For a building up to 40 stories, the proportion of core walls needed to surround the service areas naturally satisfies lateral and gravity requirements with wall thicknesses typically not exceeding 600 mm (24 in). As heights are increased, these core walls typically increase in plan area (for instance, the amount of elevators is greater); however, they are not sufficient to provide efficient resistance to drift. Therefore, combining these cores with perimeter frames through rigid diaphragms increases lateral resistance by sharing applied forces. Typically, the shear wall core dominates behavior at

Core-PT Frame,
500 West Monroe, Chicago, IL

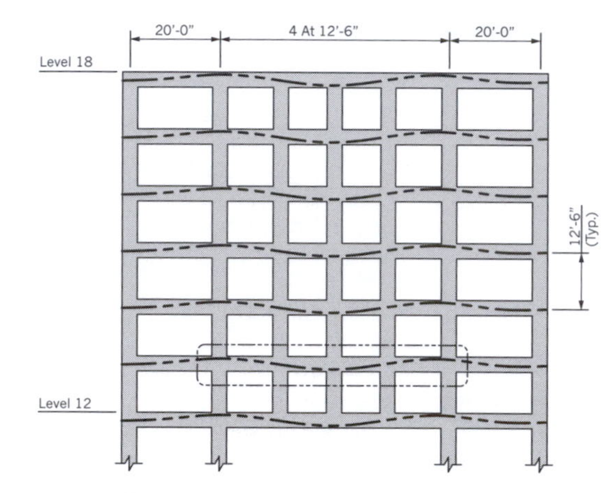

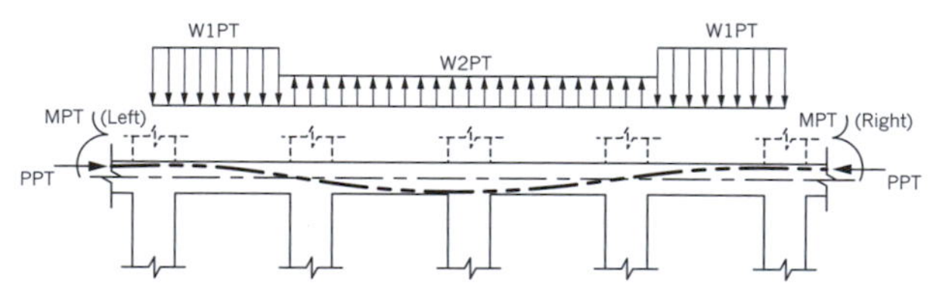

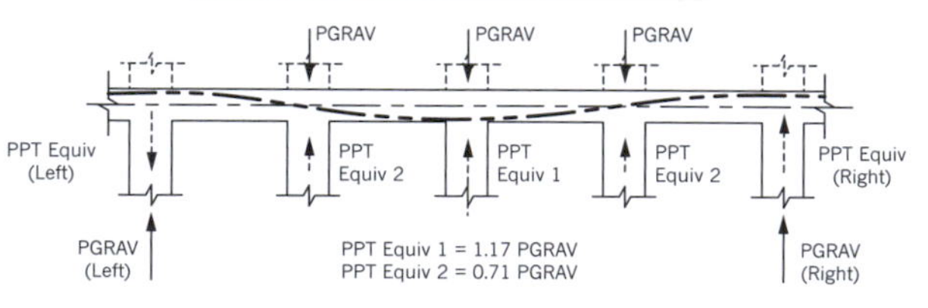

Core Wall, NBC Tower at Cityfront Center, Chicago, IL

Typical Post-Tensioned Frame Details,
500 West Monroe, Chicago, IL

the base of the building with the frames restricting the core wall cantilever behavior by acting to pull the core back at the top of the building. Managing the placement of gravity loads on frames can further increase the frame's efficiency by using gravity to counteract uplift forces. Post-tensioned concrete frames within the perimeter of 500 West Monroe in Chicago were used to redistribute axial loads, provide additional clear spans between columns, and increase structural efficiency.

9.3 THE INFINITY COLUMN

Dr. Fazlur Khan understood the limitation of conceiving a tall building as a tube, with solid but thin walls. Introduction of openings for windows was a must. However, he discovered that the placement and portioning of openings could still lead to an efficient structure. He transformed this idealistic concept into a constructible, affordable system by developing closely spaced columns and beams to form a rectilinear grid. Simply introducing diagonal members into the tubular frame achieved greater efficiency with height. This system was conceived for buildings consisting of all-steel and all-concrete.

The Jinao Tower in Nanjing, China, combines the use of local materials (reinforced concrete) and local labor to minimize cost for what

Concrete Tube-in-Tube,
Chestnut-Dewitt Tower, Chicago, IL

Concrete Braced Tubular Frame,
Onterie Center, Chicago, IL

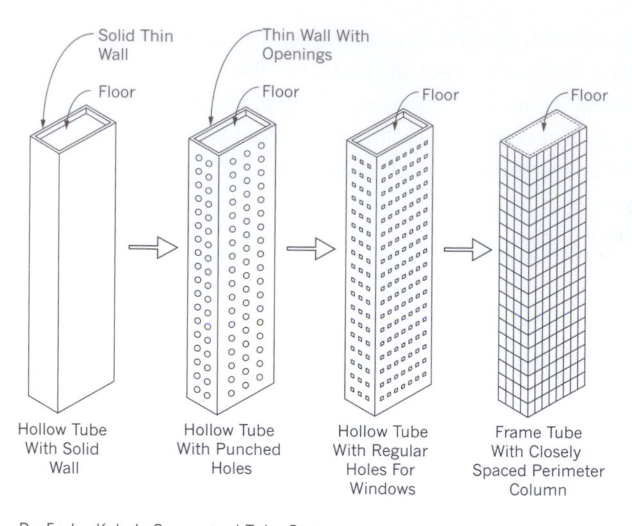

Dr. Fazlur Kahn's Conceptual Tube Systems

The Infinity Column, Kingtown International Center, Nanjing, China

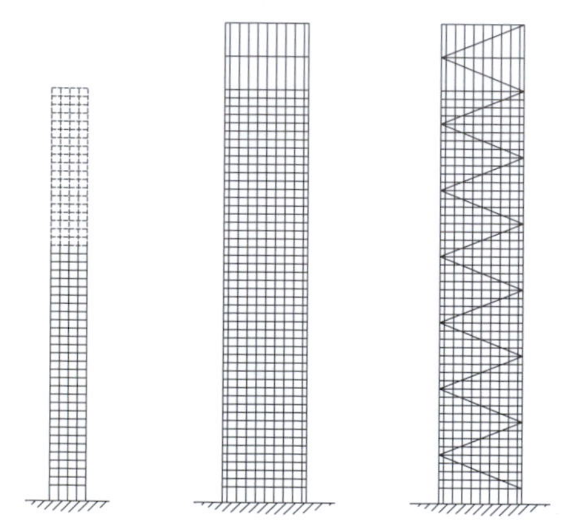

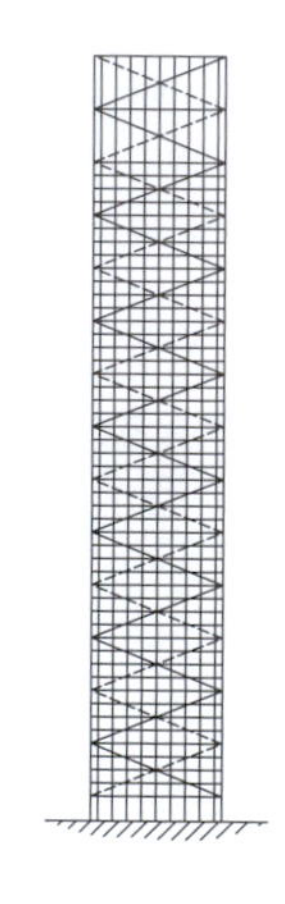

(left to right) Inner Tube Frame, Exterior Tube Frame, Exterior Tube w/Diagonals, Exterior Tube w/Diagonals Beyond, The Infinity Column Concept, Exterior Bracing Under Construction, Kingtown International Center, Nanjing, China, Exterior Bracing Under Construction

would otherwise be a conventional tube-in-tube structure. By introducing a diagonal steel member on each façade with a primary connection every four stories on lower floors and every five stories on upper floors, 45% of the rebar and concrete required for the lateral system could be eliminated resulting in a 20% decrease in material overall.

9.4 GROWTH PATTERNS

The superframe concept was developed for the initial concepts of the 137-story Columbus Center in New York. A diagonal frame wrapped angular form was designed to confuse the wind. While SOM developed the competition scheme for the China World Trade Center, Beijing, interest in bamboo as an architectural form led to the discovery of the properties of bamboo and how they might relate to the extreme high-rise. The natural formation of bamboo reveals unique structural characteristics. Long, narrow bamboo stems provide support for large foliage during its growing life while providing strong and predictable support for human-made structures after harvesting. Even when subjected to tsunamis, bamboo behaves effectively and efficiently to lateral loads exhibiting the genius of natural structural properties and geometric proportioning. The nodes or diaphragms as seen in rings over the height of the culm or stack are not evenly spaced—closer at the base, further apart through the mid-height, and close again near the top. These diaphragm locations are not random and can be predicted mathematically; they are positioned to prevent buckling of the thin bamboo walls when subjected to gravity and lateral loads. This growth pattern is common to all bamboo. The wall thicknesses and diameter of the culm can be similarly calculated. They are also proportioned to prevent buckling of the culm.

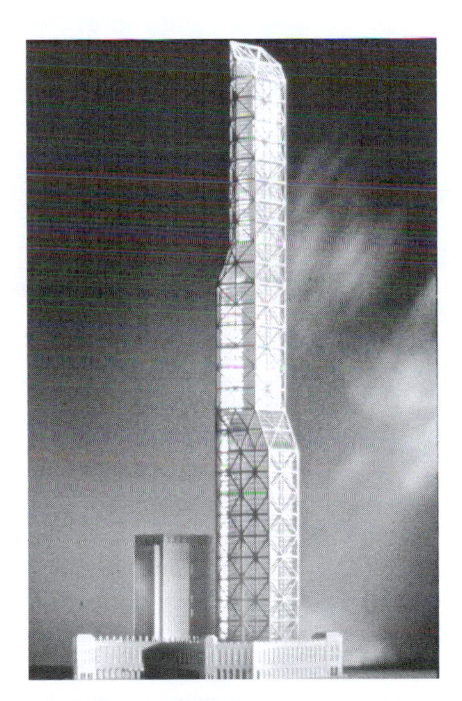

NY Coliseum at Columbus Center, Concept Model, New York, NY

Growth Pattern—Cabbage

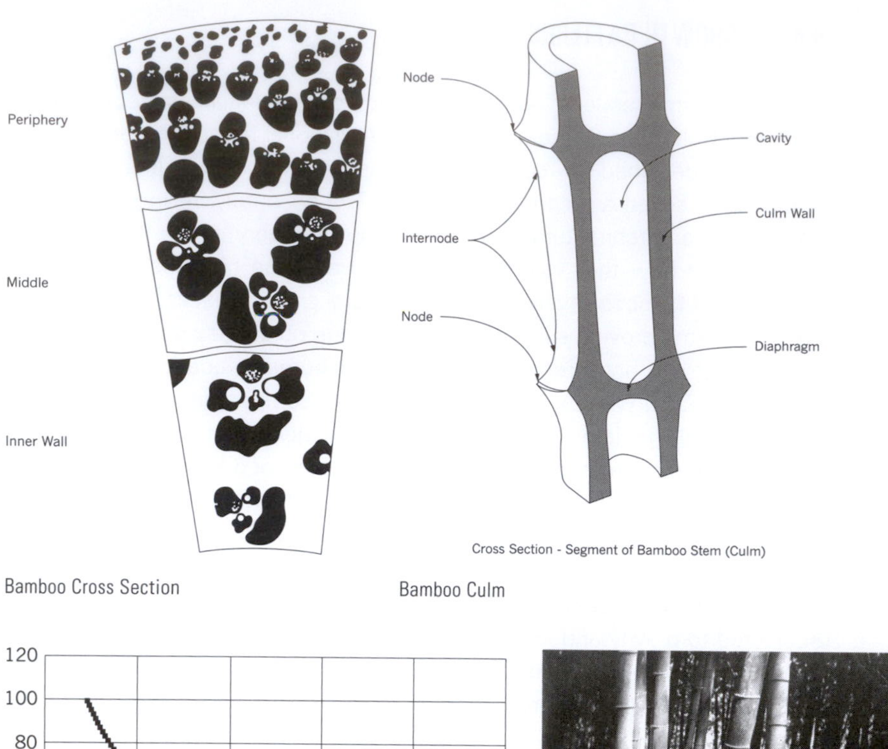

Bamboo Cross Section

Bamboo Culm

Bamboo Node Diameter vs. Height

Detail Image of Bamboo

Bamboo consists of a culm, or stem, comprising nodes and internodes. Nodes mark the location of diaphragms and provide the location for new growth. A slight change in diameter exists at node locations. Internodes exist between nodes. Internodes are hollow creating an inner cavity surrounded by a culm wall. Material in the culm is located at the farthest point from the stem's neutral axis, providing greatest bending resistance, allowing gravity loads to exist only in the outside skin which impedes uplift due to lateral loads and minimizes overall weight. The cellular structure of the bamboo wall reveals tighter cellular density near the outer surface of the wall and less density near the inner wall, again reinforcing the idea of maximum material efficiency when subjected to bending loads.

The geometric characteristics of bamboo are applied to the structural systems of the China World Trade Center Tower Competition submission. The tower is divided into eight segments along its height. The structural demand from lateral load is highest at the base of the culm (or tower) therefore internode heights are smaller compared to the mid-height. Smaller spacing increases moment capacity and buckling resistance. Beyond the mid-height of the culm (or tower) the heights of the internodes decrease proportionally with the diaphragm diameter. Thus, the form of the culm (tower) responds to structural demands due to lateral loads. Geometric relationships such as length, culm diameter, and wall thickness of many bamboo species have been previously discussed by Janssen (1991). Equations 1 through 4 define the bamboo form as discussed by Janssen (1991). Coefficients are an average of several cited species.

Internode number

$$x_n = n * \frac{100}{N} \tag{1}$$

Internode length

$$y_{n1} = 25.13 + 4.8080x_n - 0.0774_n^2 \text{ (below mid-height)} \tag{2a}$$

$$y_{n2} = 178.84 - 2.3927x_n + 0.0068x_n^2 \text{ (above mid-height)} \tag{2b}$$

Internode diameter

$$d_{n1} = 9.75 - 0.212x_n + 0.016x_n^2 \text{ (below mid-height)} \tag{3a}$$

$$d_{n2} = 178.84 - 2.3927x_n + 0.0068x_n^2 \text{ (above mid-height)} \tag{3b}$$

Wall thickness

$$t = 35 + 0.0181(x_n - 35)^{1.9} \tag{4}$$

Here x_n is the internode number, n is a shaping parameter specified by the architectural design team to be 80 based on number of floors; N is the height of the structure (320 m); y_n is the internode length; d_n is the internode diameter; t is the wall thickness. For the internode length and diameter a nonlinear relationship is observed by the transition from y_{n1} to y_{n2} and d_{n1} to d_{n2} at x_n. Thus, two polynomial equations are provided.

The relationships are shared among inner and outer structural systems. The outer structural system follows the internode length (Equation 2) with respect to mega-brace heights and mimics the culm wall fibers. The inner structural system also follows specified bamboo characteristics. Outriggers are taken as the "diaphragm" in bamboo since outriggers tie perimeter structural systems in a similar manor to diaphragms in bamboo. Internode lengths are largest at mid-height and smallest at the base and top. Diaphragm diameter is also varied over the height at outrigger levels as

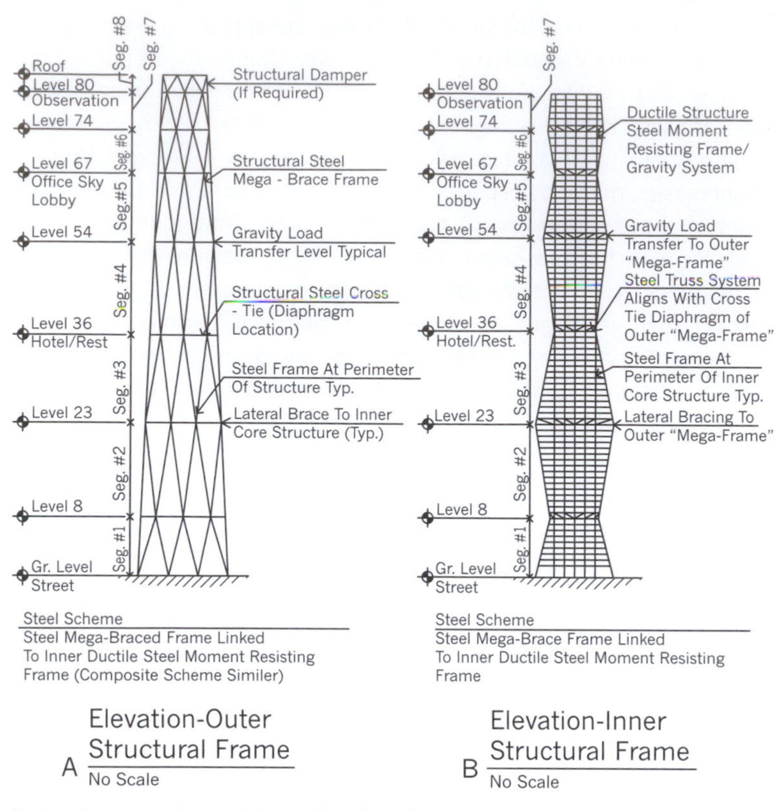

Bamboo Concepts—Structural System Elevations, China World Trade Center Competition, Beijing, China

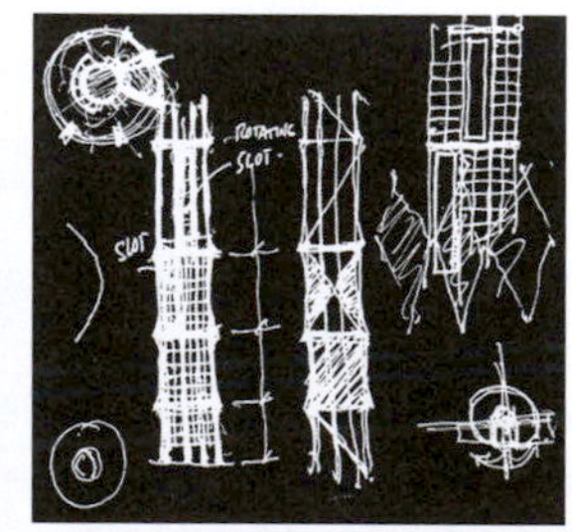

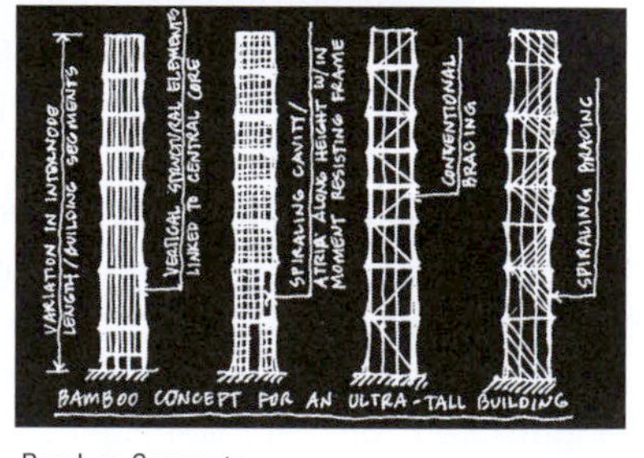

Bamboo Concepts,
China World Trade Center Competition, Beijing, China

specified by Equation 3. Finally, member sizes are proportioned to follow the wall thickness relationship shown in Equation 4.

All equations that define the diaphragm locations, diameter, and wall thickness are based on a quadratic formulation. When plotting the required diameter of the culm against height (with the relationships of diaphragms and wall thicknesses similar), it mimics the bending loading diagram of a cantilever subjected to uniform lateral loads—the engineering theory is the same for bamboo and other cantilevered structures.

9.5 THE STAYED MAST

Interconnecting a central core with perimeter mega-columns or frames provides an excellent solution to the tall tower. Building use is maximized with disruption to interior spaces limited to local areas within the structure. Outrigger trusses or walls act as levers to prop central cores. The core can be

Jin Mao Tower,
Shanghai, China

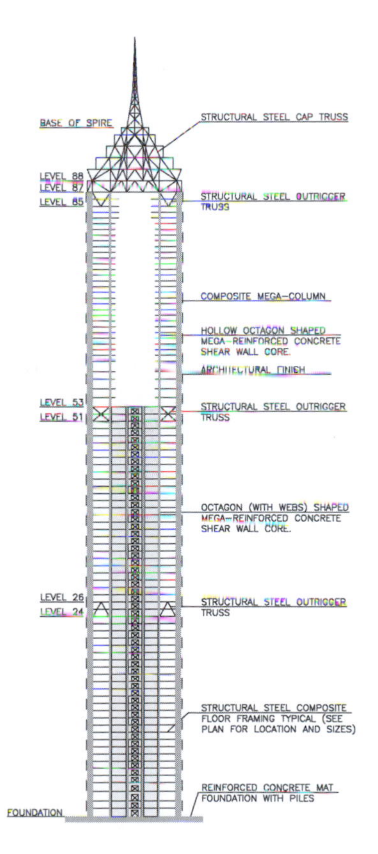

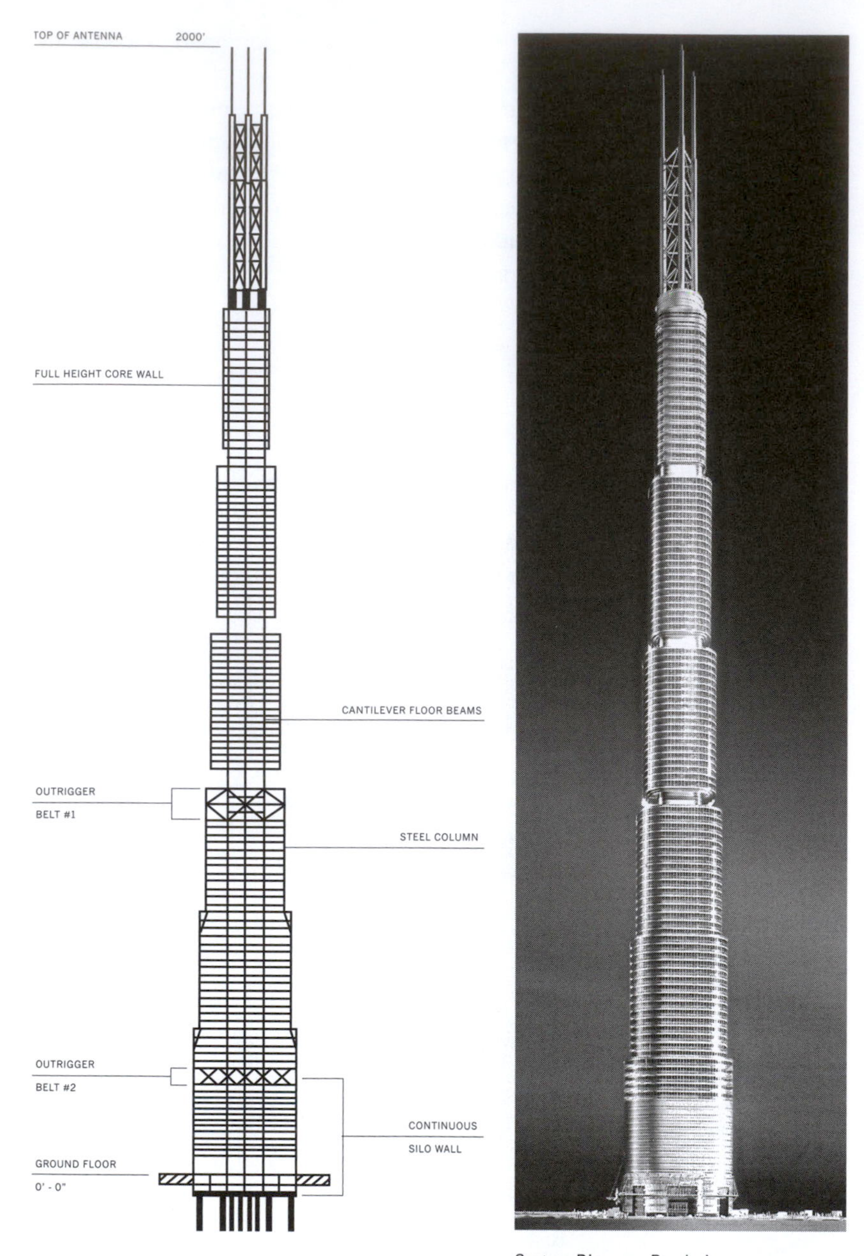

System Diagram, Rendering,
7 South Dearborn, Chicago, IL

considered a slender mast, stayed by the levers and perimeter columns. These levers develop the entire depth of the structure and are usually located at two or three locations within the height of the structure. Outrigger trusses or walls usually can be placed in mechanical areas within the tower. This system allows for heights of 365 m (1200 ft) and above.

9.6 THE BUTTRESSED CORE

At 828 m (2716 ft) the Burj Khalifa is currently the tallest building in the world. The building is not only unique because of its scale, but because the height was achieved largely with the use of concrete and the strategic position of material. Based on the concept of a desert flower, the tower incrementally steps back with height and utilizes a tripod-shaped form where a strong central core anchors three building wings. The structure is inherently stable because the strong central core anchors the three wings and each wing is buttressed to the other two. The building resists torsion through the central core while the wings provide shear and bending resistance due to an increased plan moment of inertia.

Desert Flower Inspiration for the Burj Khalifa

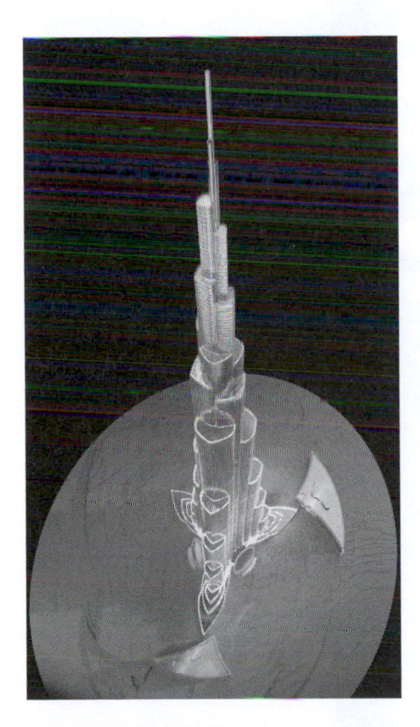

Model of the Burj Khalifa

The Burj Khalifa, Dubai, UAE

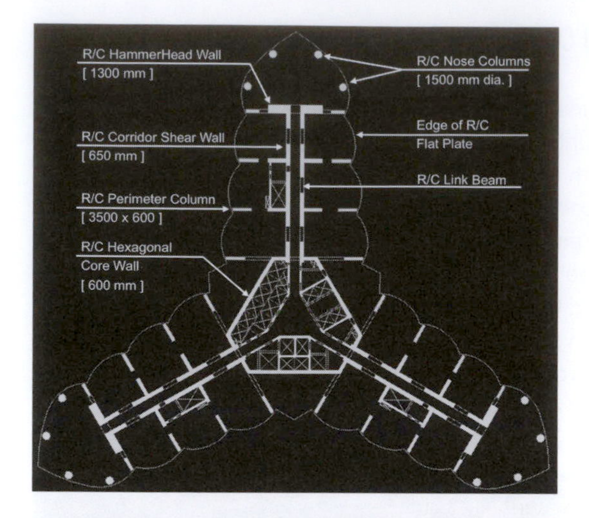

Structural Floor Plan

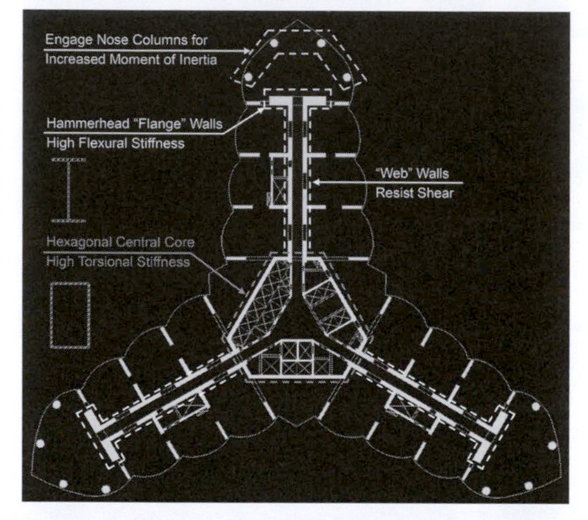

Lateral System Description

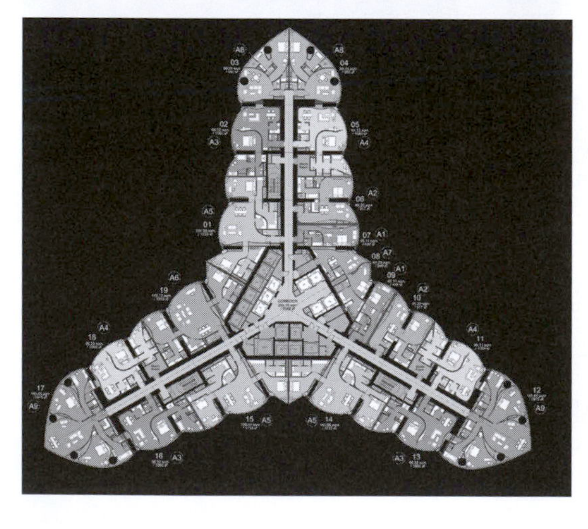

Architectural Floor Plan

9.7 THE PERFECT TUBE

Dr. Khan discovered that bundling tubular frames decreased shear lag. Shear lag is the inability of axial loads to flow around the cross-section of a tube when subjected to lateral load. Bundling the individual tubular frames not only decreases shear lag, but also increases efficiency. Efficiency is the ratio of axial column deformation to total deformation (axial, bending, and shear) in the tower. Khan sought solutions to the perfect tube and was able to increase the efficiency of the tubular frame from 61% to 78% (considering the geometry of the Sears Tower) by bundling the tubes, although he never achieved 100% efficiency.

The mesh tube conceived for the Jinling Hotel, Nanjing, China incorporates a fine diagonal mesh of structure at the perimeter of the building, eliminating any significant local shear or bending deformations of vertical members, resulting in an essentially 100% efficient structure, a structure that might lead us to expanding further the height limits of the extreme high-rise.

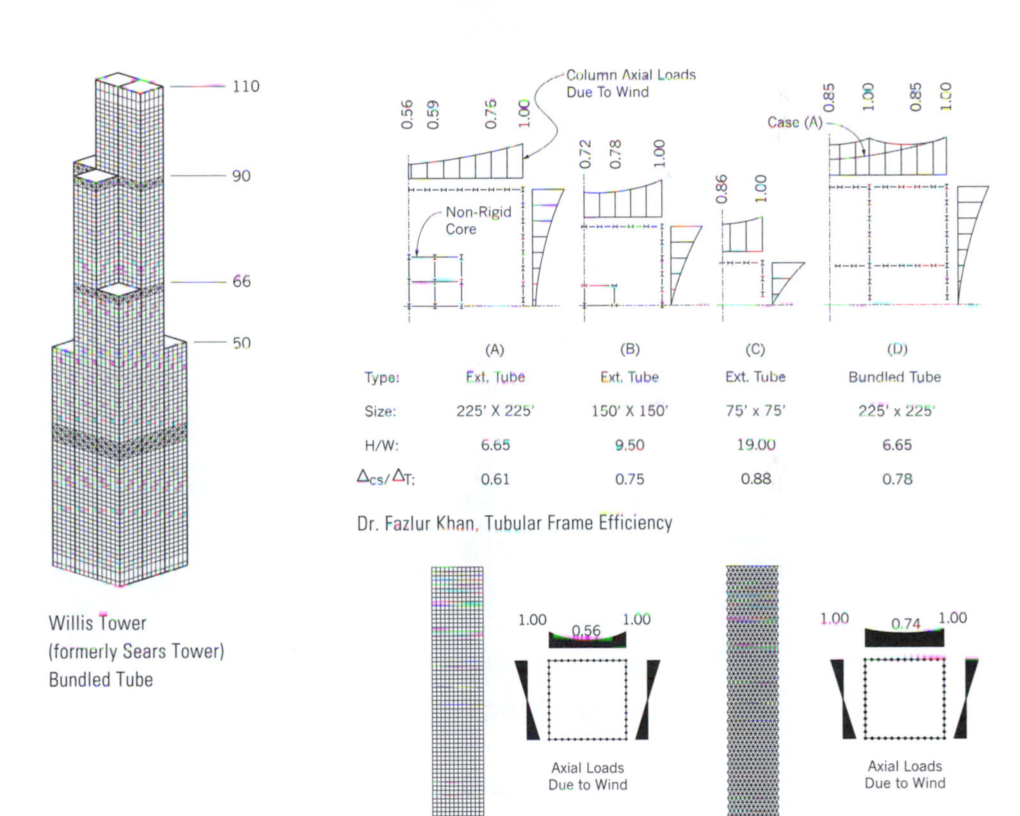

Dr. Fazlur Khan, Tubular Frame Efficiency

Willis Tower (formerly Sears Tower) Bundled Tube

(from left to right) Conventional Tube Frame, Diagonal Mesh Tube Frame

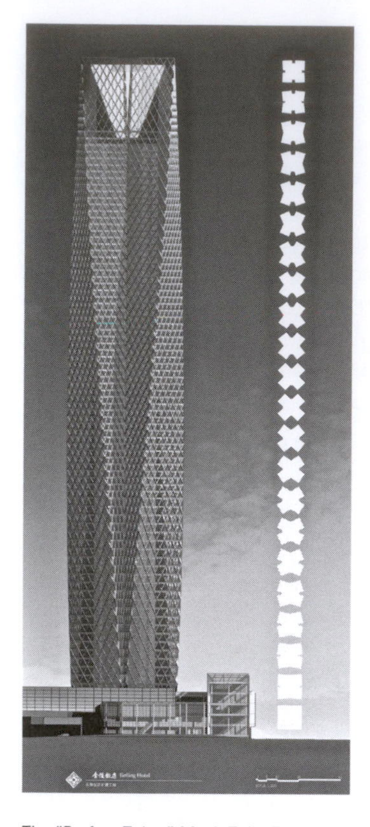

The "Perfect Tube," Mesh Tube Frame, Jinling Tower, Nanjing, China

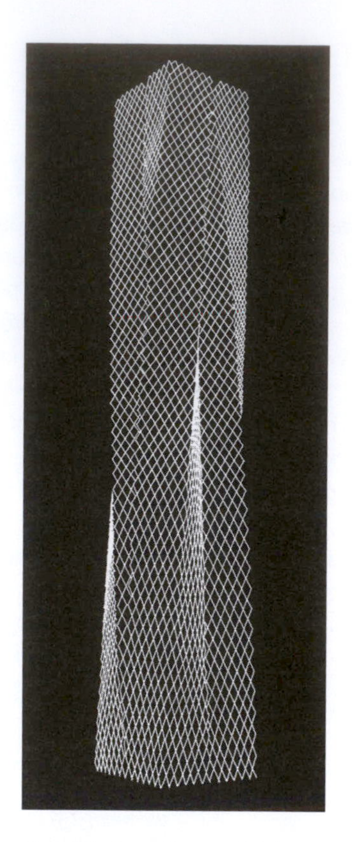

Mesh-Tube Frame Concept Sketch

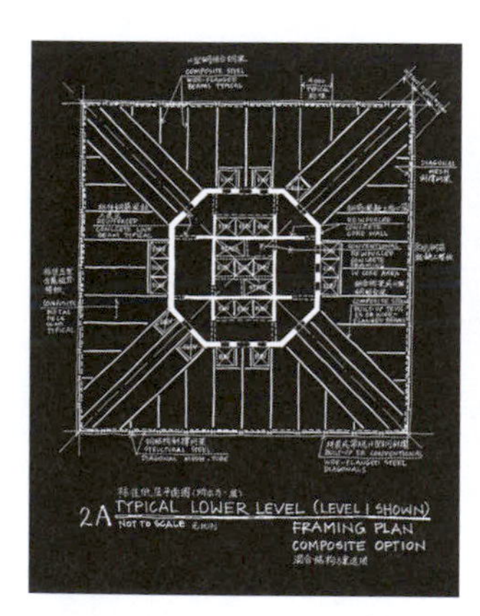

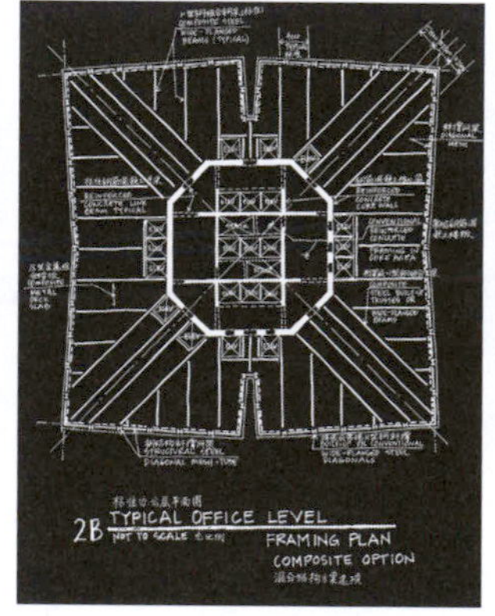

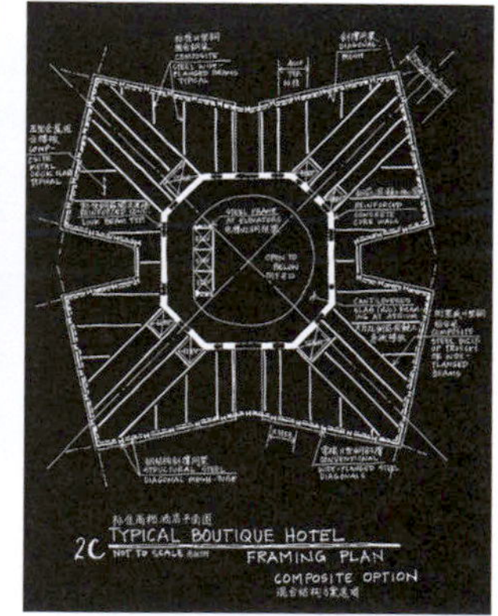

Structural System Plans, Jinling Tower, Nanjing, China

9.8 THE LOGARITHMIC SPIRAL

The logarithmic spiral—found in forms ranging from shells, seeds, and plants to spider webs, hurricanes, and galaxies can be interpreted and applied to the ultra-tall towers, scientifically mimicking natural force flows of a cantilevered structure to its foundation.

The proposed design for the Transbay Transit Tower Competition in San Francisco California is inspired by natural forms. These mathematically derived forms define systems that are safe, sustainable, cost-effective to construct, and provide optimal performance in seismic events.

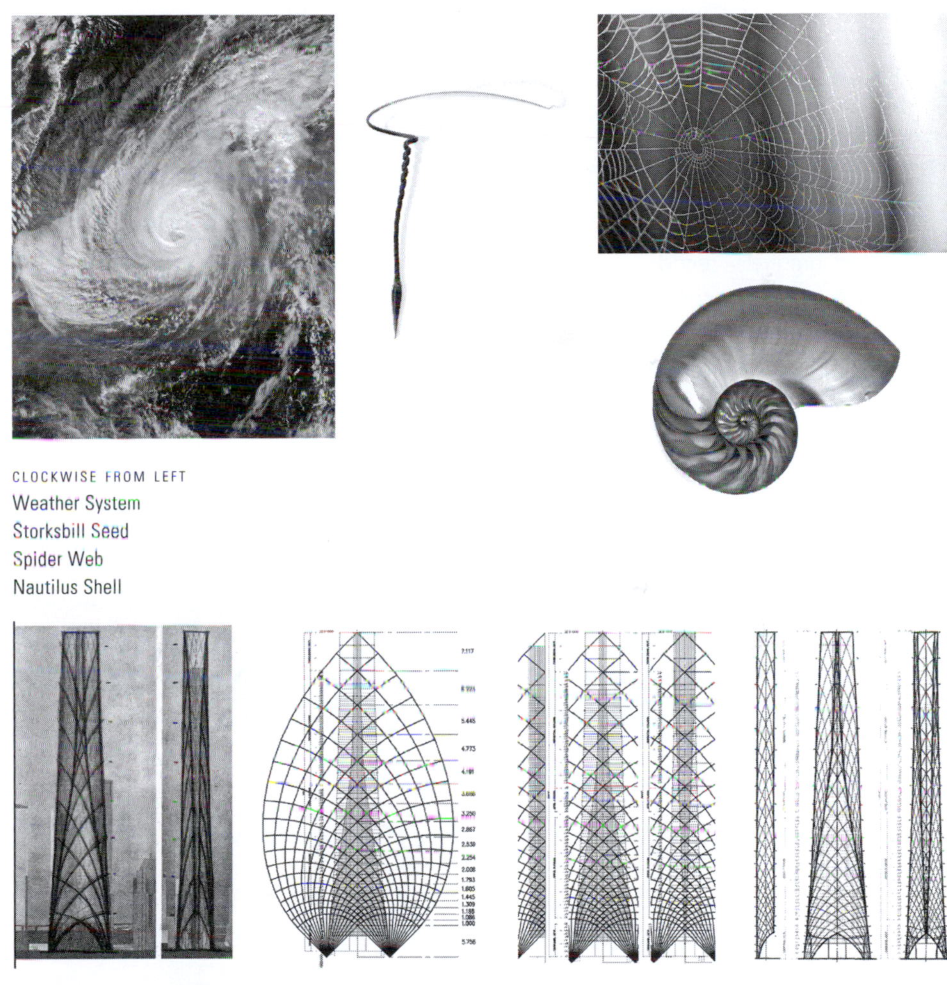

CLOCKWISE FROM LEFT
Weather System
Storksbill Seed
Spider Web
Nautilus Shell

Michell Truss Diagram Applied to the Tower Elevations, Transbay Tower Competition, San Francisco, CA

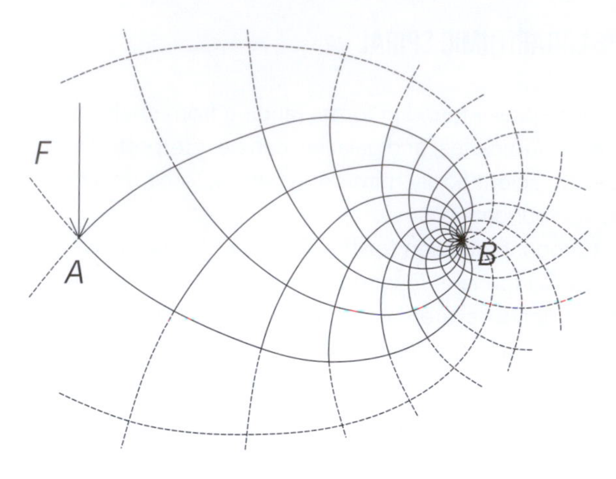

Michell Truss Diagram, A single force *F* is applied at *A*, and acting at right angles to line *AB*, is balanced by an equal and opposite force and a couple, of moment *F* x *AB*, applied at *B*. The minimum frame is formed of two similar equiangular spirals having their origin at *B* and intersection orthogonally at *A*, together with all other spirals orthogonal to these and enclosed between them

The spiral inherent in these natural forms traverses around a fixed center and gradually recedes from the center. Engineer Anthony Michell captured this behavior through his research in the early 1900s by describing the radiating lines of a pure cantilever, where force flow lines of equivalent constant stress result in specific spacing and orientations from the fixed support to the tip of the cantilever. The result is the most efficient cantilever system with the least material.

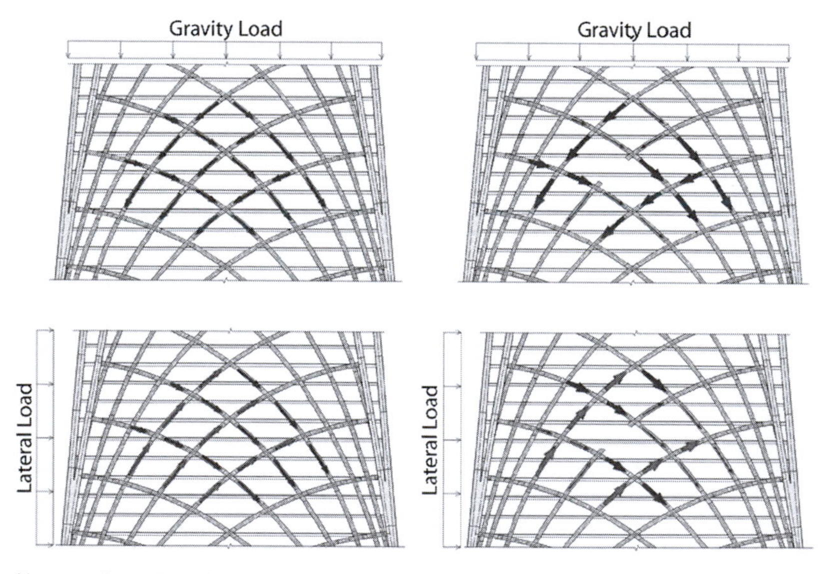

Alternate Load Path Force Flow Diagrams,
Transbay Tower Competition, San Francisco, CA

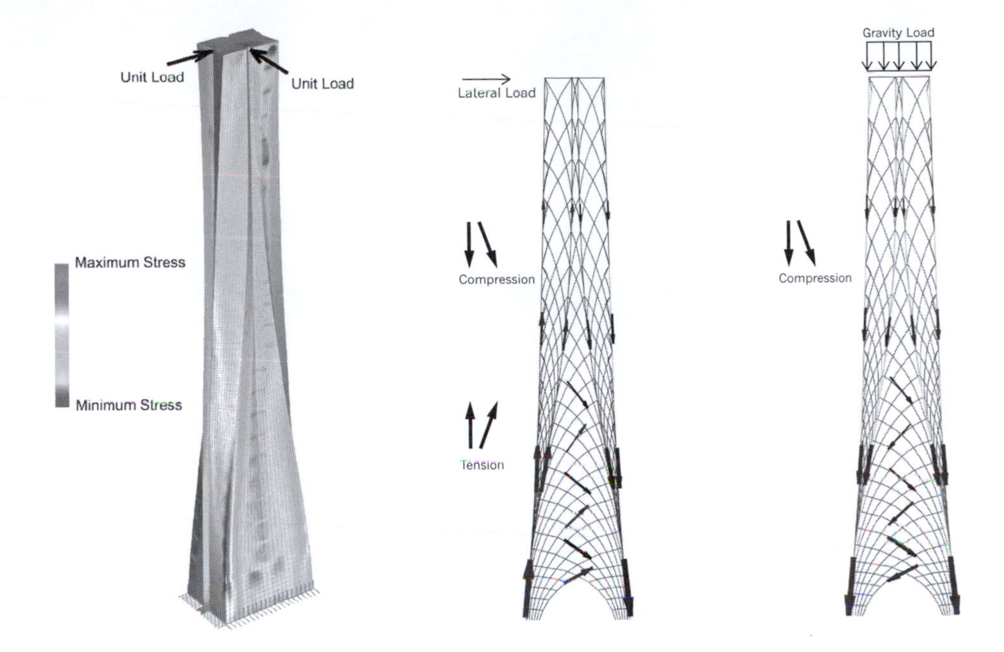

Initial Finite Element Analysis

Force Flow Diagram—Lateral Load

Force Flow Diagram—Gravity Load

The Michell Truss Diagram is mathematically interpreted and overlaid on the tower form defining an optimal perimeter bracing configuration. The structure at the base of the tower is designed to accommodate a "gateway" to the adjacent Terminal Building. The structural bracing responds to the openings in the structure, where demand is least, while mimicking gravity and lateral load force flows.

The exterior shell of the tower is robust by design, much like the spider web. It is able to self-heal in the unlikely event that a member is violated by a fire or other catastrophic event. Forces would flow to neighboring members down through the structure and ultimately into the foundations as shown. Should the structure's perimeter come under attack, the exterior composite (steel combined with concrete) bracing acts to disrupt or "shred" projectiles.

CHAPTER 10

MECHANISMS

BEYOND LIFE SAFETY CONSIDERATIONS, increasing a building's service life is paramount in regions of high seismic risk. With the implementation of scientific structural devices and systems, building services, contents, and economic investments can be protected. Building codes specifying equivalent static analyses in many cases lead to conventional designs including members and joints that have limited ductility and have questionable economic value following a major seismic event.

What if structures were designed to behave dynamically, moving freely at times, dissipating energy, protecting life safety, protecting investments, and allowing structures to remain elastic after a severe earthquake achieving the highest level of structural sustainability? What if these structures looked directly to nature for their mathematical derivations? What if the solutions to superior performance used conventional building materials? These solutions could provide a scientific response without great expense and construction complexity while increase a building structure's life cycle in regions of high seismic risk.

10.1 UNNATURAL BEHAVIOR

The 1995 Kobe Earthquake, or the Great Hanshin Earthquake as it has been formally named, was an important catalyst for considering building structures' long-term performance and life cycle. Structures in Kobe exhibited unnatural behavior; many exhibited poor ductility and many collapsed during strong ground shaking. Mid-height and ground story collapses of structures were common due to large changes in stiffness and partial height use of structural steel within reinforced concrete columns. Reinforced concrete structures lacked confinement of vertical reinforcing steel. Because of this unnatural performance of buildings, 107,000 buildings were damaged and 56,000

FACING PAGE
Transbay Tower Competition Rendering,
San Francisco, CA

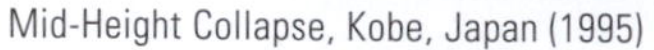

Mid-Height Collapse, Kobe, Japan (1995)

Mid-Height Column Failure, Kobe, Japan (1995)

collapsed or were heavily damaged. Some 300,000 people were left initially homeless and 5,000 people lost their lives. The total cost of the damage was well over US$100 billion (1995 US dollars).

The solution to earthquake damage is to increase behavior predictability in very unpredictable events. The natural, unpredictable event of an earthquake cannot be changed; however, forcing the structural system to behave in a predictable manner can be achieved. The solution is one that lies in elasticity of materials, which can be closely predicted, and in the passive dissipation of energy along with the inherent, internal damping of structural systems.

10.2 CONVENTIONAL BEAM-TO-COLUMN TESTS

The mid-1980s brought the conclusion of many steel beam-to-column moment connection tests. These pre-Northridge studies investigated frame connections, including the included fully welded top and bottom beam flanges and conventional shear tab web connections. The tests focused on these economical connections that incorporated typically available steel column and beam wide-flanged sections.

The tests compared the behavior of the connections without any column reinforcement, with flange continuity plates, and web doubler plates. Each connection was subjected to seven cycles of motion and was tested to at least 2% rotation to simulate expected rotation/drift in actual frame structures. Small rotations illustrated stress concentrations in beam and column flanges at complete penetration welded joints. Shear yielding in column webs could also be seen at small rotations. As larger rotations were imposed and connections were subjected to further cyclic motion, high material stress concentrations turned into fractures. Typically, flange welds failed by either failure of the weld itself or laminar tearing of the beam flange. Continuity plates provided better results, typically protecting column–beam flange connections;

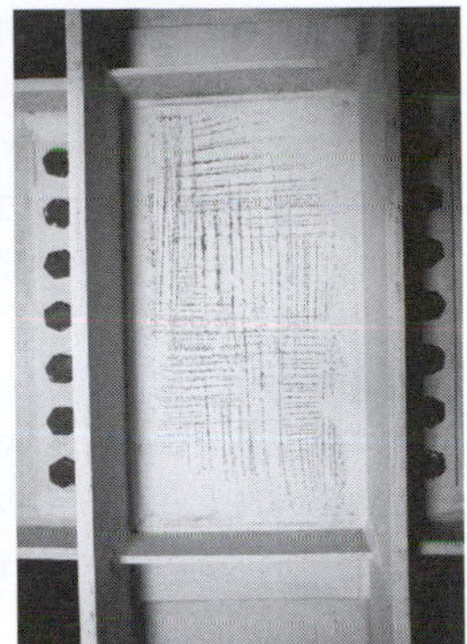

Beam-to-Column Joint Testing, Lehigh University, Bethlehem, PA

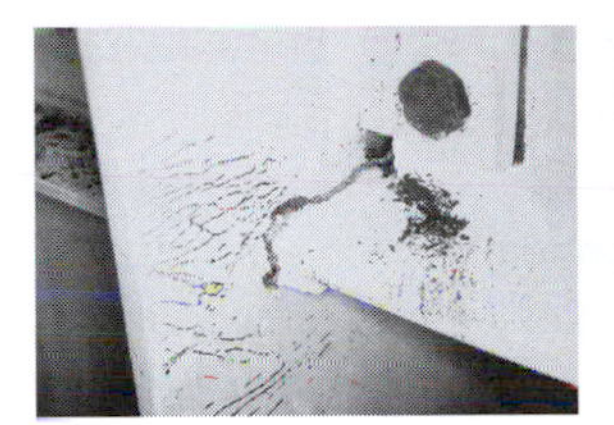

Fractured joint

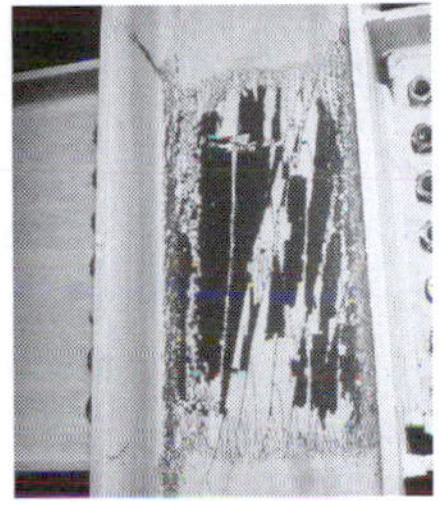

High Panel Zone
Stress Concentrations

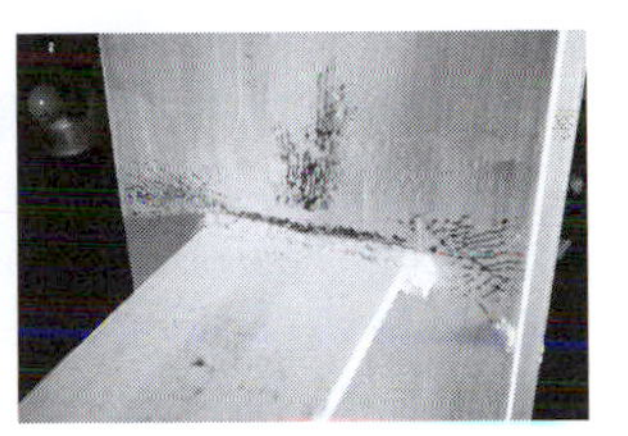

High Column Flange
Stress Concentration

however, premature flange fractures were observed. Doubler plates protected the column webs but did little to reduce stress concentrations and eventual fracture of flange welds. These plates also severely limited rotation capabilities of the joint. In some tests flange welds were removed after fracture, re-welded, inspected and tested. These connections were further tested and in some cases met the required overall rotation and drift requirements. In all cases, connections generally remained intact; beam and columns did maintain connectivity through the bolted web connections and the other full penetration welds that remained after the initial fracture.

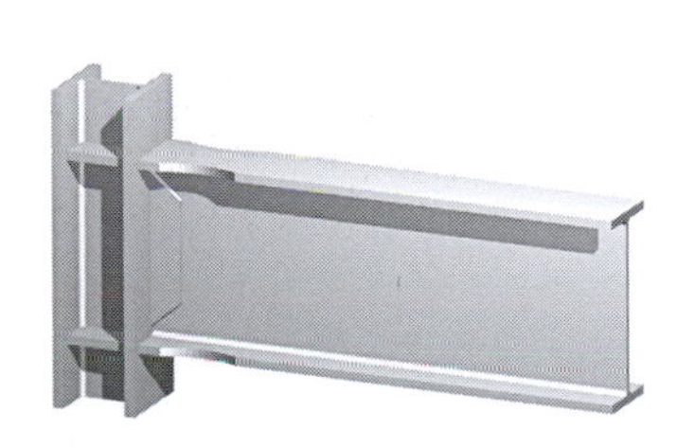

Post-Northridge Reduced Beam Section (RBS) or "Dogbone" Connection

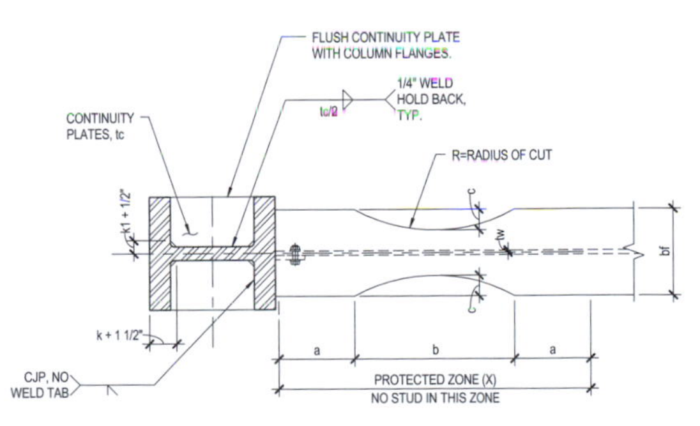

RBS Working Drawing Detail

The 1994 Northridge Earthquake in southern California proved what was found through the connection testing in the mid-1980s. Steel moment frames experienced failure of welded flange connections, but generally remained intact following the ground shaking. Life safety was protected; however, the economic loss was significant. Structural engineers were concerned about the effects of future events with questionable connections. Retrofits were made through a very difficult process of gaining access to connections through building finishes, while owners wished to keep buildings occupied.

The reduced beam section or dogbone connection was developed as an economical solution to the joint while protecting the column and ensuring ductility.

10.3 WOOD DOWELS AND STEEL PINS

Jesuit churches, built in the early 1900s, on the Island of Chiloe off the coast of Chile have performed well over their life when subjected to strong ground motion. The 1960 Chilean earthquake, perhaps the strongest earthquake ever recorded at 9.5 on the Richter Scale, did not cause major damage to these structures. Witnesses of this event describe waveforms in large open fields with amplitudes of over 6 feet. Street lights and poles used for telephone lines moved with large amplitudes and in some cases were observed to be almost parallel with the ground.

These churches have a common structural characteristic. Wood dowels from the indigenous Alerce tree were typically used to connect structural wood members with no "fixed joint" mechanical fasteners. Wood dowels were driven through openings in the structural wood members so that the dowels were snug tight and "fixed" during typical service conditions with an ability to rotate

(top) Wood Dowels

(left) Chilean Church Using Wood Dowel Connection Joints

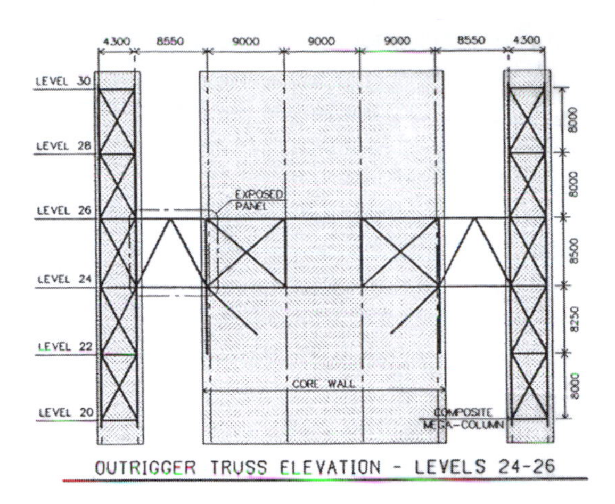

Outrigger Truss Elevation, Jin Mao Tower, Shanghai, China

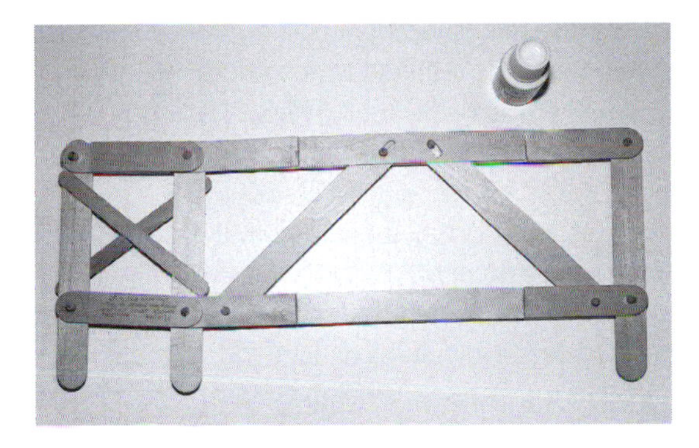

Outrigger Truss System Popsicle Stick Model

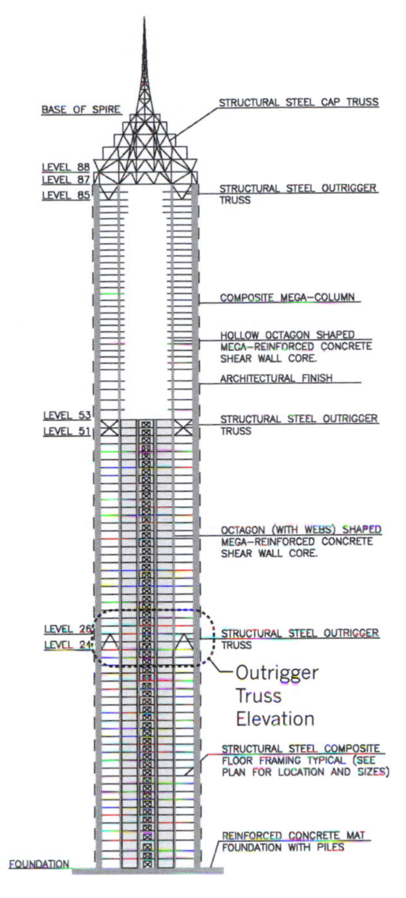

Jin Mao Tower Structural System Elevation

Jin Mao Tower Outrigger Truss Construction Photograph, Shanghai, China

during extreme loading conditions. When the joints rotated, energy was dissipated with the structures softening. With this softening the structures period lengthened, and as a result attracted less force from the ground motion.

Large diameter steel pins proved useful in the design of key structural elements of the Jin Mao Tower, Shanghai, in the People's Republic of China. The structural system concept for the 421 m (1380 ft) tall, 88-story tower included lateral load-resisting outrigger trusses that interconnected the central reinforced concrete core and eight perimeter composite mega-columns. This interconnection occurred at three two-story segments between Levels 24 and 26, 51 and 53, and 85 and 87. Relative creep, shrinkage, and elastic shortening between core and the composite mega-columns was significant; if this behavior were not managed, the forces developed in the outrigger trusses would be large enough to overload the members causing significant oversizing, yielding, and potential failure. If fully connected at the time of erection, up to 45 mm (1.75 in) of relative displacement was expected between the core and composite columns with only 16 mm (0.625 in) of relative displacement expected after 120 days. By delaying the fully bolted connections for 120 days and including the pins to allow for free movement, the structure was only required to accommodate 16 mm (0.625 in) of relative displacement after the bolts were installed. To manage these forces and reduce the amount of structural material required for the members and connections, large-diameter steel pins were incorporated into the truss system. Based on basic concepts of statics, these pins allow the trusses to behave as freely moving mechanisms prior to making final connections. This mechanism concept allowed the trusses to be constructed during normal steel erection procedures and allowed final bolted connections to be installed after a significant amount of relative movement had already taken

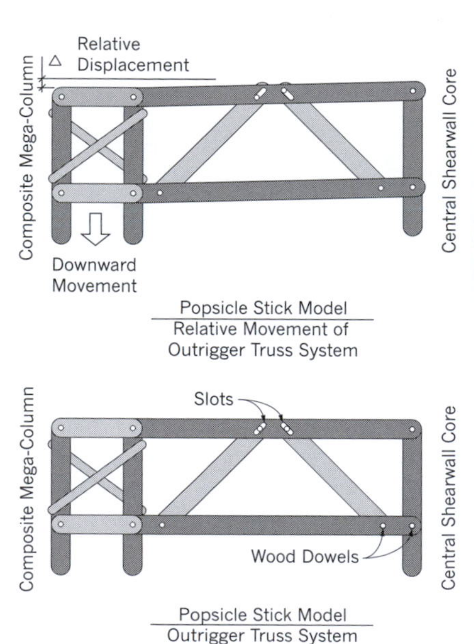

Pin in Outrigger Truss Joint,
Jin Mao Tower Construction Photograph

Outrigger Truss System, Popsicle Stick Model,
Jin Mao Tower, Shanghai, China

place. Because diagonal members were used within the trusses, large slots were incorporated into end connections to allow movement to occur. These slots were the key component to all free movement of the system.

A simple model made up of tongue depressors, wood dowels, and popsicle sticks was used to study the behavior. This conceptual model was the basis for the structural solution considering building materials, then analyzed, and developed into full working drawings.

10.4 PINNED JOINTS

The Jin Mao pinned-truss concept, based on the fundamentals of statics and behavior, led to the development of a series of structural systems that will behave predictably in an extreme seismic event. The systems are appropriate for building and other structures with varying heights and geometries allowing systems to remain elastic and dissipate energy while protecting economic investments.

Structural steel end connections of members contain pins or bolts and are installed to a well-calibrated torque by applying compression of joints through the pin or bolt tension. Faying surfaces are treated with an unsophisticated slip-type material such as brass, bronze, cast iron, aluminum or hard composite

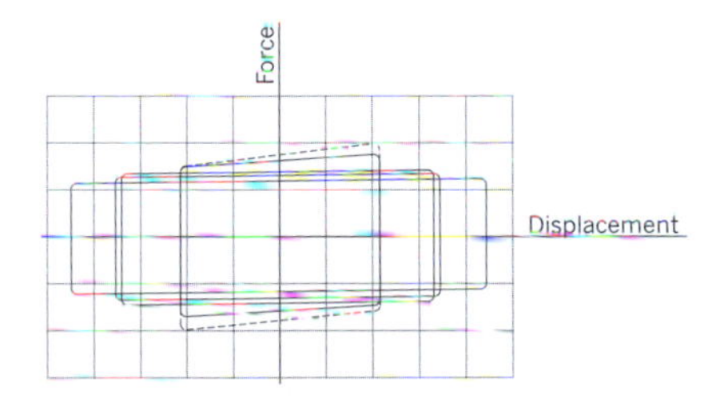

Representative Hysteresis Loop

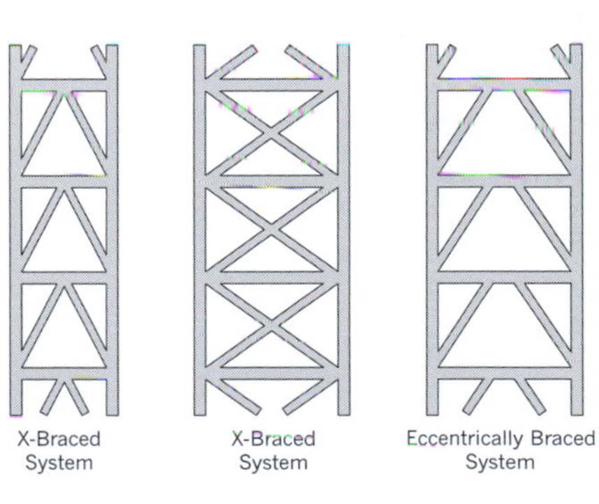

Truss System Elevations

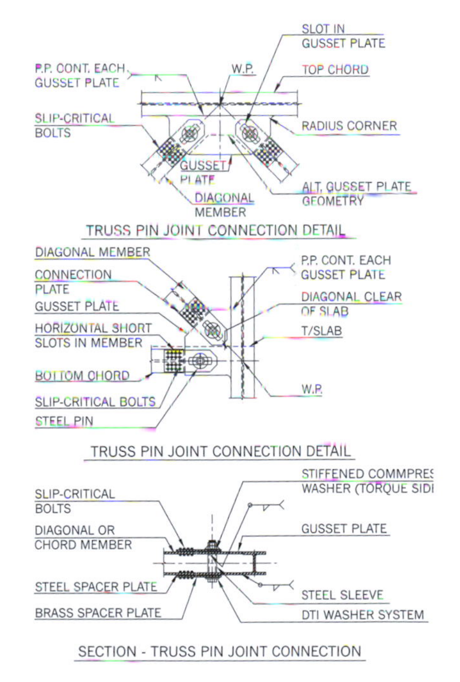

Pin-Connection Details

materials with a well-defined coefficient of friction, allowing for significant movement capabilities after the threshold of movement and without significant loss of bolt or pin tension. The material combination not only provides well-defined load-displacement characteristics, but also excellent cyclic behavioral attributes. Structural steel surfaces can be sandblasted or cleaned with mill scale present to minimize galling or binding. A force vs. displacement hysteresis curve of a combination of steel and brass illustrates the stable behavior.

Pins, installed into the end connections of trussed building frames used in buildings subjected to extreme seismic events, provide fixed joints for the typical service life of the structure and are allowed to slip during extreme events.

Circular holes are used in horizontal truss members for both connection plates and members. Circular holes are used in diagonal members with long-slotted holes used in connection plates. Slots are designed and installed in the direction of force. The length of the hole is dependent on the predicted drift and the dynamic characteristics of the structure when subjected to ground motion. A well-defined faying surface is used between connection plates and members. The material sandwiched between the steel surfaces is a shim or pad form (may be applied directly to the steel surfaces) and is stringently protected after application, during shipment, and during erection. The faying surface material is used in both slotted and circular connection joints. Pins are torqued in slip-defined joints at slotted connections of diagonal members. The size of the pin and the amount of torque are directly related to the coefficient of friction of the material being compressed between steel plates and the expected force that will initiate joint motion. Special Belleville washers are used to help maintain bolt tension after slip has occurred in slotted joints. In addition, direct tension indicators (DTIs) are used under the head (non-torquing side) to ensure proper pin or bolt tension. Structural members are designed for forces required to initiate the onset of joint motion. Members are designed to remain elastic, not allowing yielding, local or global buckling.

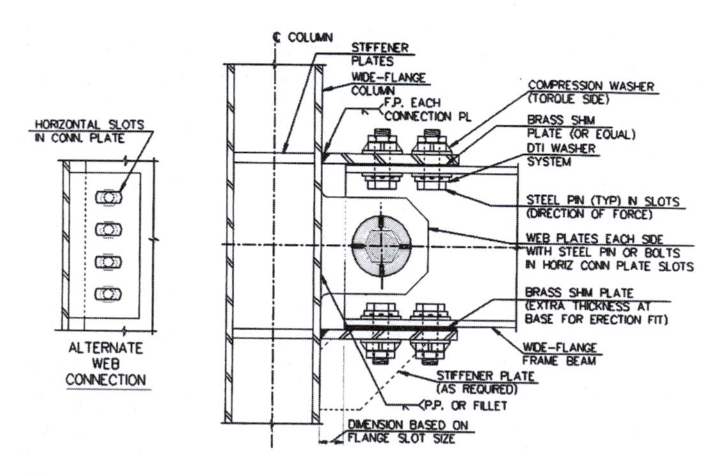

Moment Frame Beam-Column Joint Detail

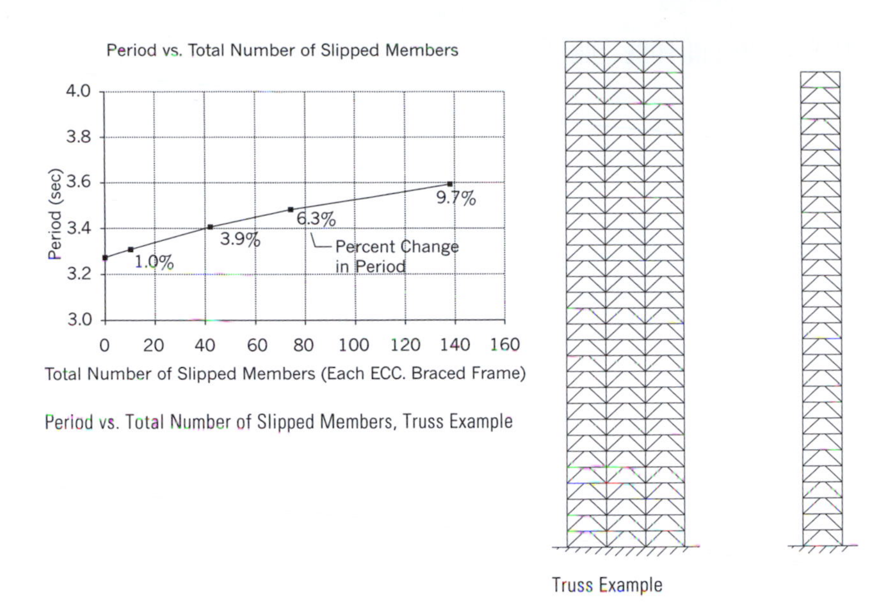

Period vs. Total Number of Slipped Members, Truss Example

Truss Example

Moment-resisting frames equally benefit when pins are introduced into end connections. Pins in long-slotted connections can be introduced into top and bottom beam flanges. The length of the beam flange slots can be predicted by evaluating the expected beam rotation and inter-story drift characteristics. A single pin in a circular hole is used in the web connection to provide the best rotational capabilities.

These connections, whether placed in braced or moment frame systems, act to lengthen the structure's period. This lengthening of the period effectively softens the structure, resulting in less force being attracted to the structure from the ground motion.

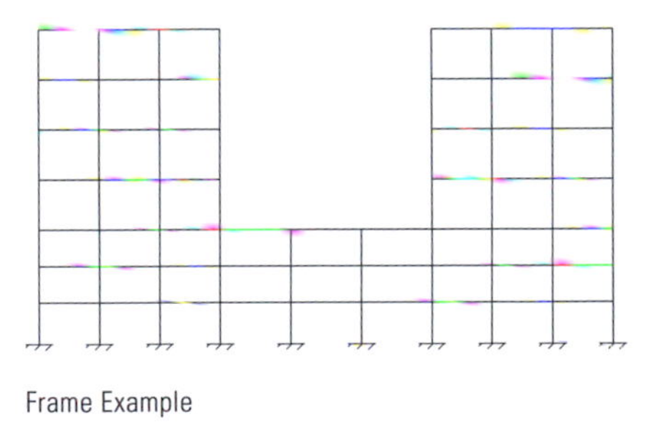

Frame Example

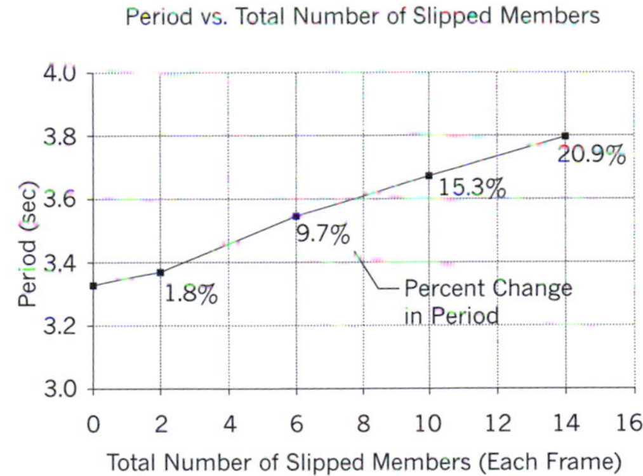

Period vs. Total Number of Slipped Members, Frame Example

10.5 THE PIN-FUSE JOINT®

If structures were capable of altering their characteristics to resist potentially destructive forces during extreme loading events without permanent deformation, their expected life cycles would significantly increase.

Structural frames previously utilizing beam-to-column moment connections that are welded with the frame beams perpendicular to the columns would become obsolete. Beams are typically connected to the face of columns rotate when subjected to racking of the building frame. These beams are designed to protect the column integrity and prevent potential collapse by plastically deforming during frame motion. This deformation, however, likely decreases post-earthquake integrity and economic viability in the process. Therefore, a new concept for this connection is warranted.

Based on the concept that structures should behave dynamically rather than statically during seismic events, the Pin-Fuse Joint® allows joints to be fixed in building frames until they are subjected to extreme loads. High-strength bolts, brass or non-metallic shims and curved steel plates create fixity with a well-defined coefficient of friction.

The Pin-Fuse Joint® allows building movement caused by a seismic event, while maintaining structural elasticity after strong ground motion. The joint introduces a circular-plated end connection for the steel beams framing into the steel or composite columns within a moment-resisting frame. Slip-critical friction-type bolts connect the curved steel end plates. A steel pin or

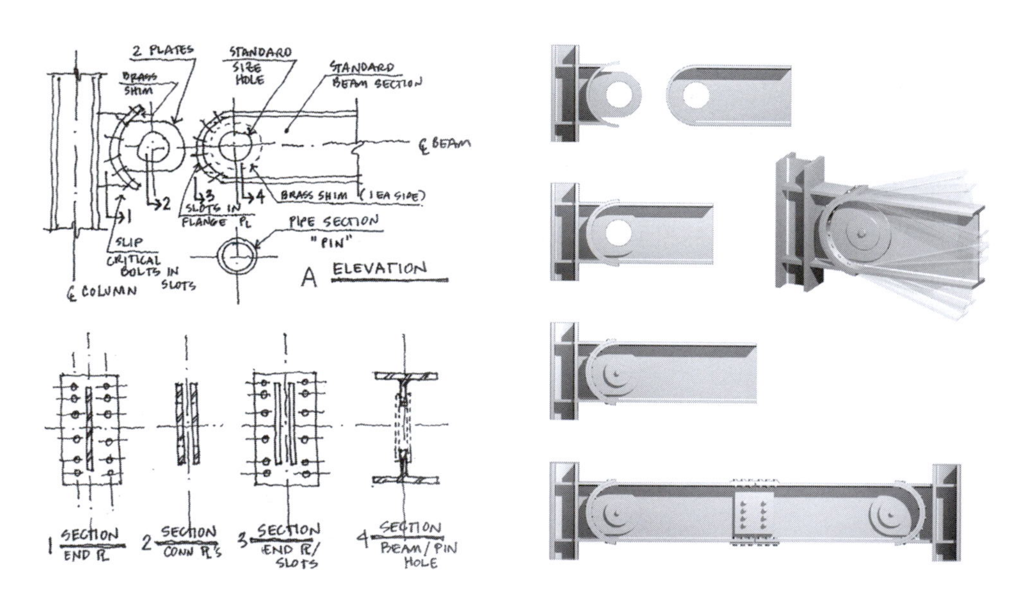

The Pin-Fuse Joint®
(Patent No. US 6,681,538 B1 & US 7,000,304)

The Pin-Fuse Joint®
(Patent No. US 6,681,538 B1 & US 7,000,304)

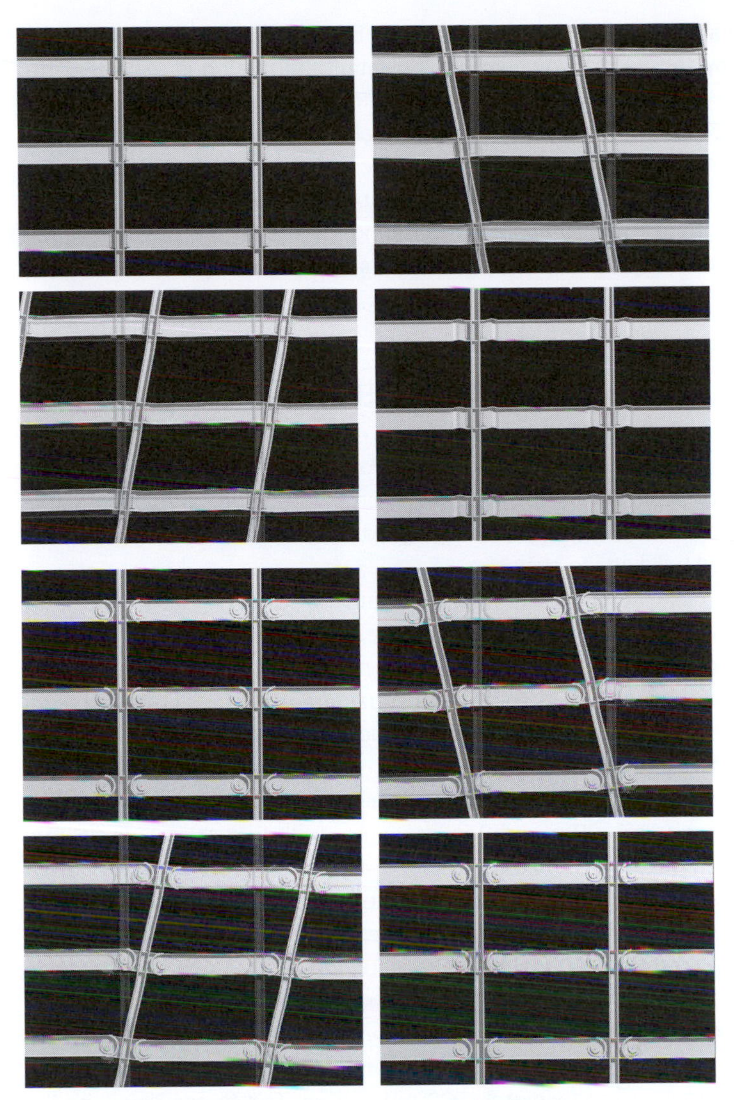

Behavior of Conventional Moment Frame versus Pin-Fuse Joint® Frame

hollow steel pipe in the center of the moment-frame beam provides a well-defined rotation point. Under typical service conditions, including wind and moderate seismic events, the joint remains fixed where applied forces do not overcome the friction resistance provided between the curved end plates. However, during an extreme event, the joint is designed to rotate around the pin joint, with the slip-critical bolts sliding in long-slotted holes in the curved end plates. With this slip, rotation is allowed, energy dissipates, and "fusing" occurs.

The rotation of the Pin-Fuse Joint® during extreme seismic events occurs sequentially in designated locations within the frame. As the slip occurs, the building frame is softened. The dynamic characteristics of the frame are altered with a lengthening of the building period, and less force is attracted by the frame; however, no inelastic deformation is realized. After the seismic event, the elastic frame finds its pre-earthquake, natural-centered position. The brass or non-metallic shim located between the curved steel plates provides a predictable coefficient of friction required to determine the onset of slip and also enables the bolts to maintain their tension and consequently apply the clamping force after the earthquake has subsided, resulting in complete pre-earthquake fixity.

10.6 MANAGING LARGE SEISMIC MOVEMENTS

The use of pinned mechanisms in structures is not limited to frame joints; such mechanisms may also be used in large structural components that assist in natural building behavior during extreme seismic events and allow for increased building life cycles.

A rocker mechanism, installed at the mid-height of the Poly Corporation Headquarters in Beijing, China, allowed the building to freely move during a significant earthquake, while providing support for the world's largest cable net and for a mid-height museum used to display some of China's most important antiquities. The cable net, 90 m (295 ft) tall and 60 m (197 ft) wide at its widest point, was conceived to provide support for the exterior wall system with minimal structure. Stainless steel cables were typically spaced at 1500 mm (59 in) on-center with vertical cables 26 mm (1 in) in diameter and horizontal cables 34 mm (1.34 in) in diameter. A V-shaped cable stayed concept was used to reduce the cable net span and reduce overall displacements when subjected to wind loads. Cables 200 mm (8 in) in diameter not only provided lateral support for the cable net, but also were used to support the mid-height museum. It was found that these large cables acted as diagonal mega-braces when the building was subjected to strong ground motion. The displacement at the top of the building relative to the mid-height was approximately 900 mm (36 in). The forces developed in the cables and connections with this level of relative displacement could not be reasonably resisted. If forces in the primary cables could be relieved during lateral motion, then cables and connections could be designed for wind and gravity loads only. A pulley, located at the top of the museum, could achieve this behavior. However, because a drum size nearly 6 m (20 ft) in diameter was required to accommodate the primary cables and was aesthetically unacceptable, an alternate idea was required.

A rocker mechanism, capable of movements in two primary directions, was introduced into the top of the museum. Pins combined with steel castings provided the reverse pulley mechanism capable of resisting the imposed loads while remaining elastic during an extreme seismic event.

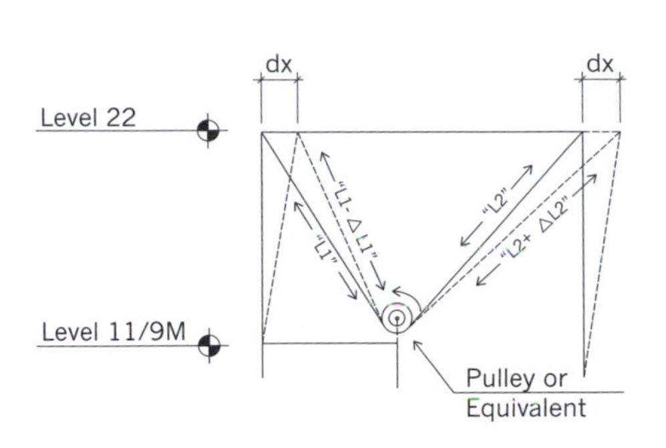

Pulley Concept, Poly Corporation Headquarters, Beijing, China

Poly Corporation Headquarters, Beijing, China

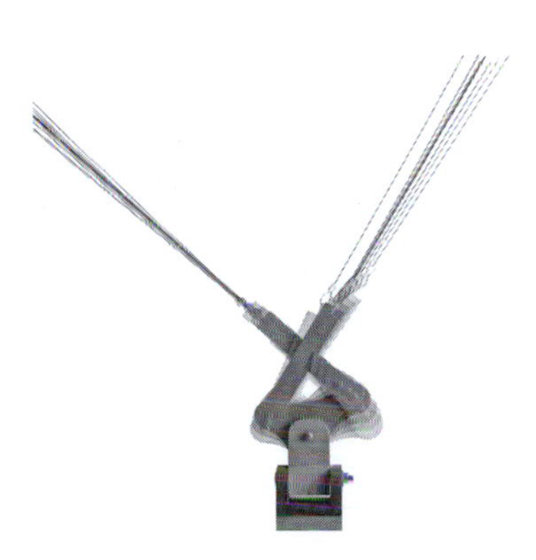

The Rocker, Poly Corporation Headquarters, Beijing, China

Study Model, Rocker Location at top of Museum Box, Poly Corporation Headquarters, Beijing, China

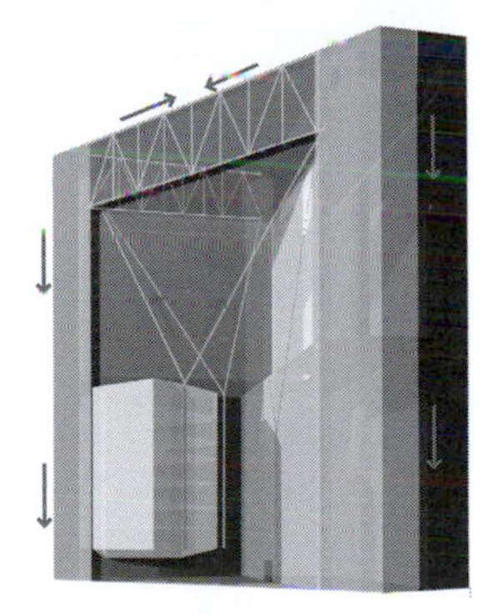

Force Flow, Poly Corporation Headquarters, Beijing, China

10.7 COMBINING NATURAL FORMS AND MECHANISMS

The combination of natural forms and mechanisms results in the minimum use of material and the best life cycle. Therefore, a dual system could be incorporated into a structure: one system controlling lateral drift and one providing fused mechanisms used to protect the structure while maintaining permanent elasticity, as illustrated in the plan for the Transbay Tower.

Transbay Tower Competition Rendering

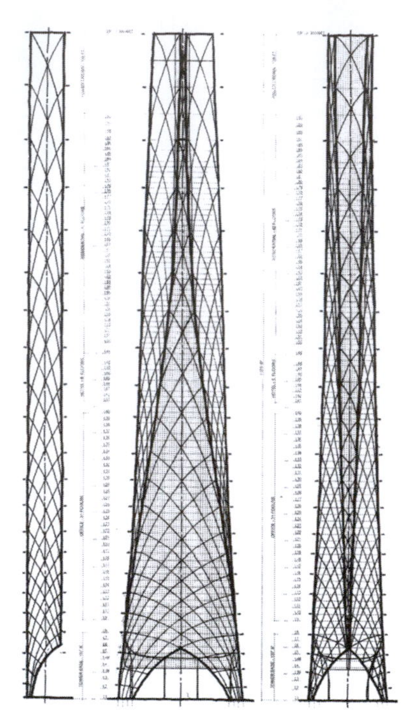

Transbay Tower Competition Elevations

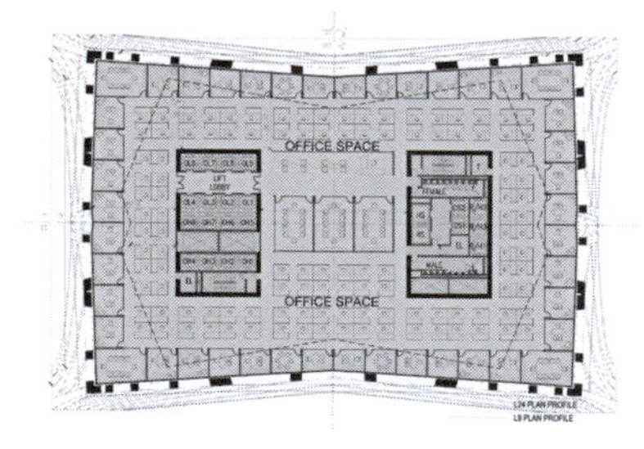

Transbay Tower Competition Floor Plan

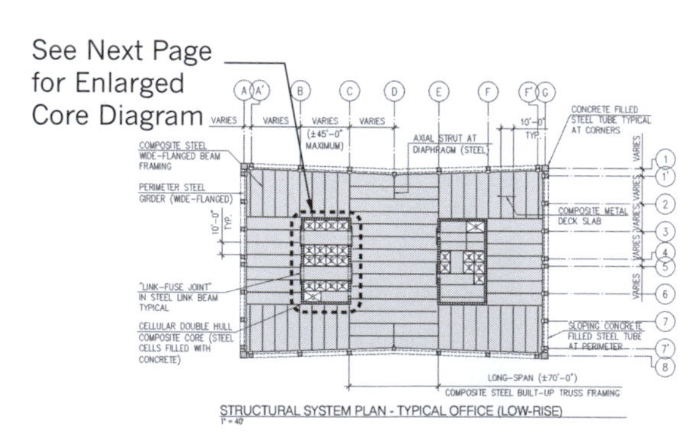

Transbay Tower Competition Structural System Plan

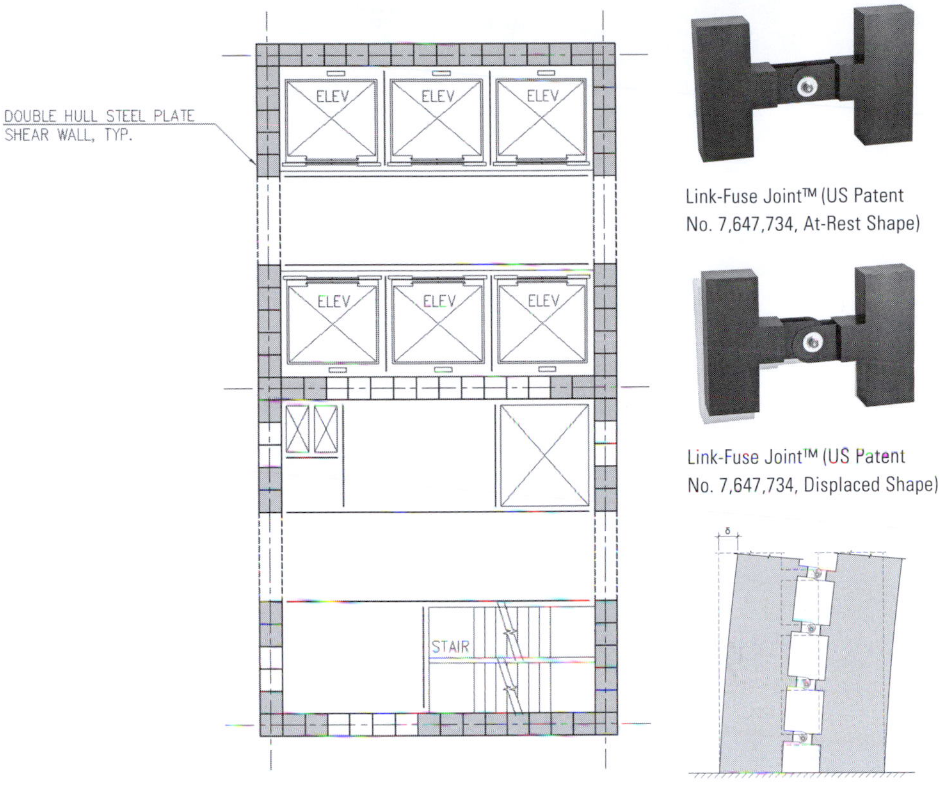

Transbay Tower Competition Enlarged Core Diagram

Link-Fuse Joint™ (US Patent No. 7,647,734, At-Rest Shape)

Link-Fuse Joint™ (US Patent No. 7,647,734, Displaced Shape)

Link-Fuse Device in Link Beams

The core of the tower is designed to protect inhabitants and guarantee a safe path of egress in the event of an emergency. The core walls are hardened with a cellular structural concept derived from maritime construction. This "double hull" wall includes a steel plate shell filled with concrete and provides an armored barrier for elevators, stairs, and primary mechanical life safety systems. This barrier also provides optimal fire resistance.

The tower is designed to resist the most extreme earthquakes and remain operational in tandem with the essential Terminal facility. The tower incorporates an innovative array of seismic fuses designed to slide during extreme seismic events. This Link-Fuse™ system allows the building core to dissipate energy at wall link locations (where openings are required to enter the core) while protecting the rest of the structure from damage. After the earthquake subsides and the building comes to rest, the fuses maintain their load carrying capacity and the building can immediately be put back into service.

The Link-Fuse Joint™, incorporated into the reinforced concrete shear wall (or steel frames) of structures, allows links that typically occur over doorways and mechanical openings to fuse during extreme seismic events. The butterfly slot pattern in the steel connection plates is clamped together,

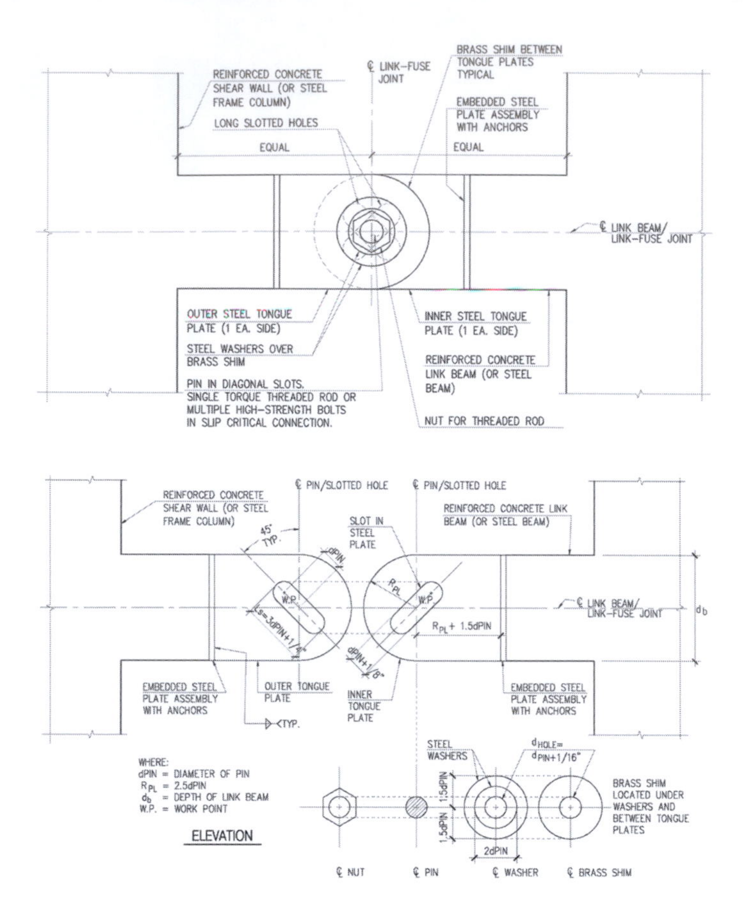

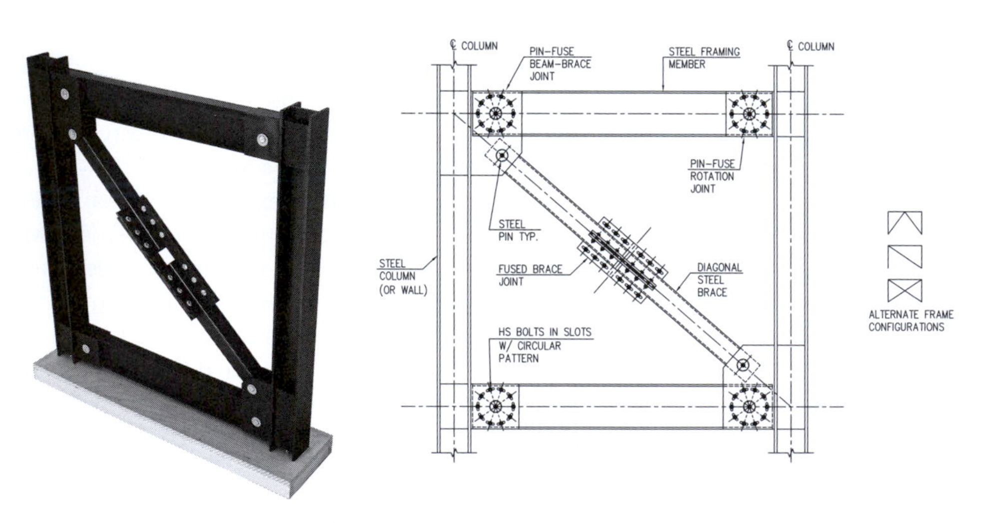

The Pin-Fuse Frame™
(US Patent No. 7,712,266)

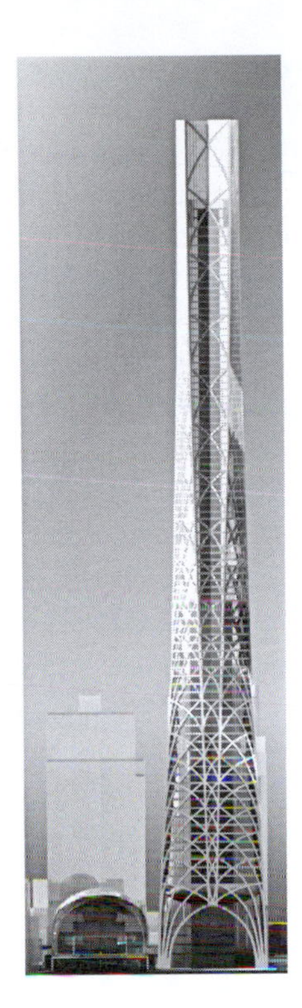

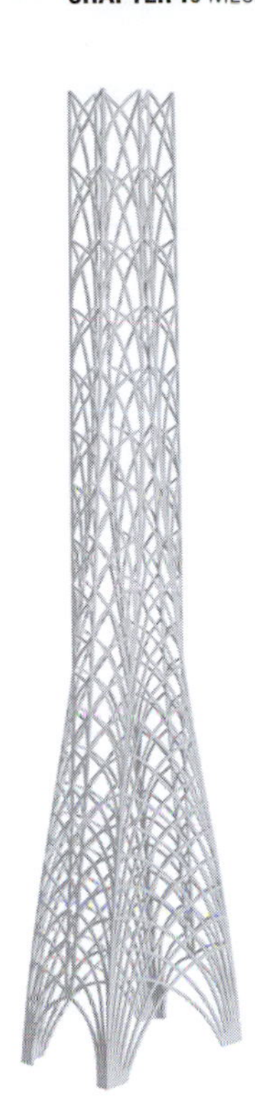

Transbay Tower Competition, San Francisco, CA

developing static friction between plates. Brass shims located between the plates create a well-defined and consistent coefficient of friction. When the demand on the links is extreme, the joints slip in any necessary vertical or horizontal direction, dissipating energy, softening the building (period lengthens), and attracting less force from the ground. After the motion of the building and the movement of the joints cease, the building and joints return to their natural at-rest position without permanent deformation. The tension in bolts used to clamp the plates together is not lost during the motion and, therefore, re-establishes the structural capacity by re-establishing the static friction within the joints. The structure is safe, its economic investment protected, and it can remain in service. In contrast, conventional reinforced concrete link beams are typically damaged and must be repaired or replaced, with the overall structure potentially deemed unfit for future service.

The Pin-Fuse Frame™, incorporated into braced frames (or between concrete shear walls) of structures allows braces to slip or "fuse" during extreme seismic events. High-strength bolts in long-slotted holes are used to clamp a sandwich of brass or non-metallic shims and steel plates (brass or shims between the plates). Brass or shims placed between steel plates creates a well-defined threshold of slip when subjected to load. When the demand on the structure is extremely high, the moment-resisting frame with its circular bolt pattern provides additional resistance. If these joints are subjected to a high level of bending moment, they too will slip, rotating around a center steel pin. The combination in behavior of the braces and horizontal moment-resisting elements fuse the structure, dissipate energy, soften the building, lengthen the period, and reduce forces attracted from the ground motion. After the event, these joints return to their full structural capacity without permanent deformation. The tension in bolts used to clamp the plates together is not lost during the motion and, therefore, re-establishes the structural capacity by re-establishing the static friction within the joints. Although conventional steel braced frames have been proven to perform with some adequacy during extreme seismic events, their integrity is questionable in repetitive cyclic motions due to premature buckling. Damage in many cases may not be repairable.

The structural topology conceived at the perimeter combined with the fused core could provide what may be the most efficient structural system for a tall building anywhere in the world, reducing material quantities required for construction and providing a truly sustainable structure designed not only to survive but also to remain in service after even the most significant natural and unnatural disasters.

Transbay Tower Competition, Tower Base, San Francisco, CA

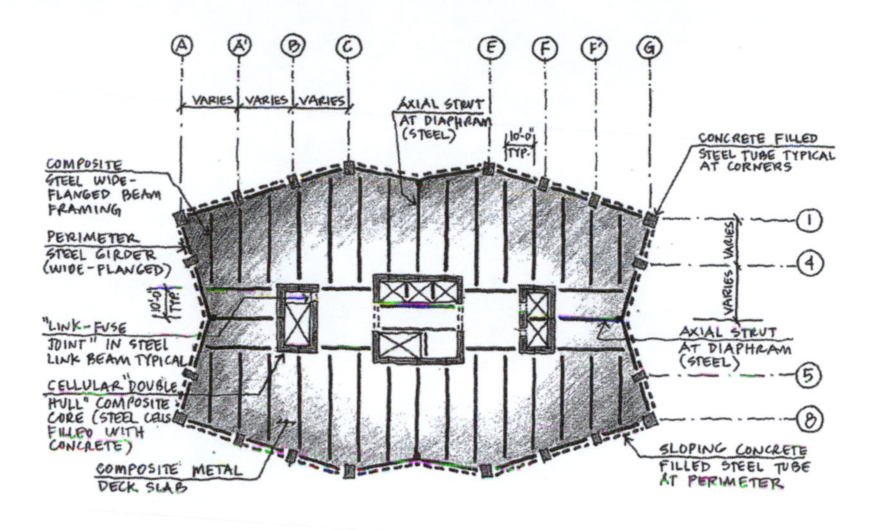

Transbay Tower Competition,
Residential Framing Plan, San Francisco, CA

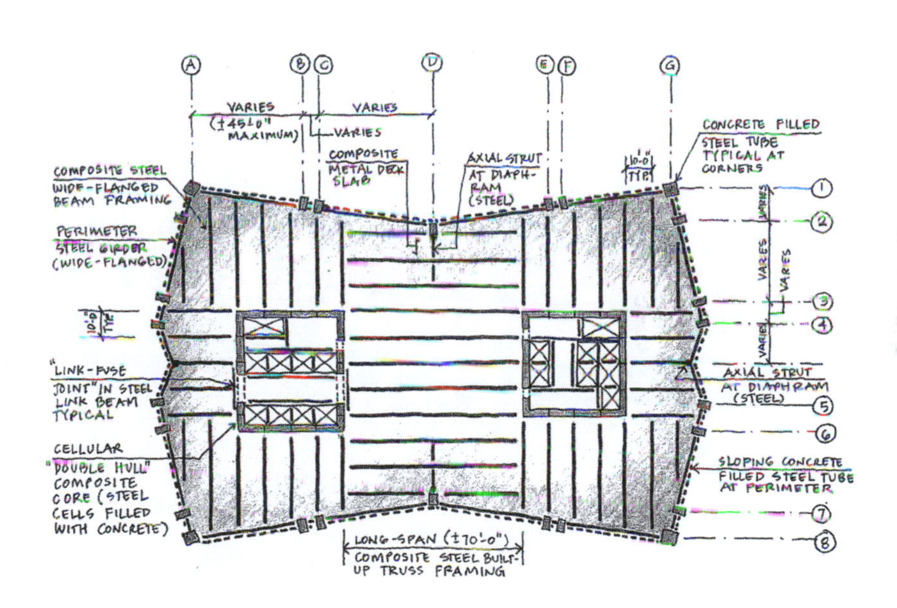

Transbay Tower Competition,
Office Framing Plan, San Francisco, CA

CHAPTER 11

PERFORMANCE

CONTRARY TO COMMON PERCEPTION, performance-based design does not necessarily lead to structures with better performance. This method of design specifically addresses seismic performance of tall buildings, including structures with long fundamental periods of vibrations, significant mass participation and lateral load response in higher modes of vibration, and a relatively high aspect ratio (slender profile). The process typically results in a better understanding of structural behavior, but does not lead to enhanced performance unless specific higher performance objectives (minimum objectives are defined in the building code) are used, including ground motion input, components, or systems. In addition, building codes have one primary goal and that is to protect the public and create structures that are life-safe. Many people have the incorrect perception that buildings designed to current code are earthquake-proof, but in fact they can sustain significant, and in some cases even irreparable, damage even when they are designed to be life-safe.

Rendering
(Structure-Core Only Lateral System),
500 Folsom Street, San Francisco, CA

FACING PAGE
Digital Model, The City
of San Francisco

Performance-based design was originally developed for structures that are an exception to the building code through an alternative, non-prescriptive approach to design with the methodology confirming code equivalence. These structures typically complied with the general requirements of the building code, with the exception of height limits described for particular seismic force-resisting systems, or for seismic force-resisting systems not specifically described in the code. For instance, concrete structures of various mass and stiffness characteristics over 49 m (160 ft) tall located in areas of high seismicity require a dual structural system composed of a shear wall core and moment-resisting frame to meet the prescriptive intent of the building code. Because of increased cost, increased construction time, and architectural impacts, many have designed and successfully built these structures without the frame by proving code equivalency through performance-based design methods.

11.1 OBJECTIVES

There are two primary objectives for performance when considering a non-prescriptive analysis and design. The first required is defined as a minimum performance objective where code equivalency is met, and the second is an enhanced performance objective where systems and components are designed to standards beyond those required by the building code.

11.1.1 Minimum Performance Objectives

11.1.1.1 Maximum Considered Earthquake (MCE)

The structure must withstand an MCE earthquake with a low probability (approximately 10%) of collapse without loss of gravity load carrying capacity, without inelastic straining of important lateral load carrying elements that would result in severe strength degradation, without excessive permanent lateral drift, and without the development of global structural instability. In addition, all elements of the structure must be designed for compatibility with the anticipated deformations of the seismic-force-resisting system. The MCE level earthquake typically has a probability of exceedance of 2% in 50 years or a return period of 2475 years.

11.1.1.2 Design Earthquake (DE)

The structure must withstand a DE earthquake also known as a Design Basis Earthquake (DBE) having an intensity 2/3 x the MCE without creating significant hazards to individual lives while assuring that non-structural components and systems remain anchored in place and building drifts are limited so that undue hazards are not created. The DE level earthquake typically has an equivalent probability of exceedance of 10% in 50 years or a return period of 475 years.

11.1.1.3 Frequent or Service Level Earthquake

The structure must withstand a relatively frequent, more moderate-intensity earthquake shaking with minimal damage. The structure must be designed to remain essentially elastic considering service level earthquake ground shaking of a 50% probability of exceedance in 30 years or a return period of 43 years.

11.1.2 Enhanced Performance Objectives

It may be desired, and it is encouraged, to design structures to achieve a performance higher than the life-safety requirements of the code. The following sections set out examples of enhanced performance objectives.

11.1.2.1 Probability of Exceedance

A lower probability of exceedance for service level or MCE level ground shaking or both can be considered in selecting ground motions. The intensity of shaking will be high resulting in designing the structure for greater seismic demand.

Under Construction (Core-Only Lateral System),
350 Mission Street, San Francisco, CA

11.1.2.2 Drift and Residual Displacement

Establishing lower limits for lateral drift and/or reduced levels of acceptable levels of cyclic straining of ductile elements will lead to enhanced performance. Limiting residual displacements will lead to a structure that can be more easily repaired after an earthquake (most primary structural components will have remained elastic).

11.1.2.3 Non-Structural Components

Non-structural components and systems could be designed for accelerations based on higher ground motion intensities or story drifts that are larger than those required by the building code.

11.1.2.4 Damage-Tolerant or Response Modification Devices

Damage-tolerant structural elements such as pin-fused seismic systems that are capable of remaining elastic during strong ground motions, or other devices that can withstand cyclic inelastic deformation or limit permanent distortion, will lead to enhanced performance. Response modification devices such as seismic isolation, energy dissipation systems such as viscous dampers, or passive and active control systems can be used to enhance performance and limit damage.

11.2 DESIGN APPROACH

Since the design approach takes exception to the building code, it is important to obtain concurrence that this approach is acceptable to the building official reviewing the project. Once that is established, the following procedure should be used:

1. Performance Objectives—either minimum or enhanced.
2. Seismicity—two levels of ground motions must be considered. The service level earthquake must consider a 2.5%-damped acceleration response spectrum having a 43-year return period and the maximum considered earthquake shaking with a 5%-damped acceleration response spectrum having a 2475-year return period.
3. Conceptual Design—structural systems and materials, as well as intended elements that will be subjected to inelastic or pseudo inelastic beforehand, must be selected.
4. Design Criteria—a design criteria that establishes the design and analysis approach with all performance objectives, system types, codes/references, and materials must be developed and submitted for review by the building official and a third party peer panel.
5. Preliminary Design—a dynamic structural analysis must be used to confirm that the design is capable of meeting the established

performance objectives. To perform this design, the structure must be developed to a level of detail where stiffness, strength, and mass are defined, as well as hysteretic properties of elements that must undergo inelastic straining due to strong ground motion. To the extent possible, the structure should be configured to include simple arrangements of structural elements with clearly defined load paths and regular structural systems. Large changes in building stiffness, building mass as well as repositioning of bracing elements over the height of the tower, column transfers, and system eccentricities should be avoided. This will limit complexity and uncertainty in the final design.

6. Service Level Evaluation—this consideration must be used to demonstrate that the structure is capable of withstanding a frequent seismic event with limited damage.
7. Maximum Considered Earthquake Evaluation—a non-linear dynamic analysis must be used to demonstrate that the structure will not collapse during this level of ground shaking.
8. Final Design—since the final design described in building codes is based on the design or design basis earthquake, 2/3 times the intensity of the MCE must be considered with all load combinations and strength/response modification factors applied.
9. Peer Review—because the design process is non-prescriptive and takes exception to one or more requirements of the building code, an independent third party peer review is required. The peer review panel typically consists of experienced engineers: one an expert in seismology, one an expert in the practice of engineering, and one an accomplished academic.

11.3 PERFORMANCE-BASED DESIGN EXAMPLE

The following is an example of performance-based design used for a 30-story office tower located in downtown San Francisco.

11.3.1 Structural System Description

The 350 Mission Street tower is 117.1 m (384 ft, 2 in) tall above grade with three basement levels below grade and a total building area of approximately 42,293 square meters (455,000 square feet). The primary proposed structural system consists of reinforced concrete from the foundation to roof. Since the building is over 73 m (240 ft), the core-only lateral system is an exception to the building code that required a shear wall core to be combined with a moment-resisting frame.

Rendering, 350 Mission Street, San Francisco, CA

11.3.1.1 Superstructure

The superstructure consists of a reinforced concrete shear wall core and perimeter gravity columns with two-way flat plate slab framing. The tower is roughly square in plan with dimensions of 38.1 m x 39.6 m (125 ft x 130 ft). The typical office floor-to-floor height is 4 m (13 ft, 2 in).

11.3.1.2 Lateral System

The lateral system consists of a reinforced concrete shear wall core only. The shear wall core has an external plan area of 13.1 m x 16.0 m (43 ft x 52 ft, 6 in) and is located around the service area of the structure, around passenger and service elevators, as well as back-of-house areas. The shear wall core extends from foundation to roof. The shear walls vary in thickness from 838 mm to 610 mm (from 33 in to 24 in) and in concrete compressive strength from 55 MPa to 41 MPa (8,000 psi to 6,000 psi). The shear wall core is interconnected with the use of ductile reinforced link beams at openings required for doorways and corridors.

11.3.1.3 Gravity System

The gravity framing system consists of a 275 mm (11 in) thick two-way flat plate post-tensioned concrete slab with a compressive strength of 34 MPa

Long-Span Post-Tensioned Flat Place Construction,
350 Mission Street, San Francisco, CA

(5,000 psi). The perimeter vertical gravity columns typically comprise conventional reinforced concrete sections varying in size from 1100 mm x 1100 mm (42 in x 42 in) to 660 mm x 660 mm (26 in x 26 in). The columns utilize concrete with compressive strengths ranging from 55 MPa to 41 MPa (8,000 psi to 6,000 psi). The tall columns at the entry lobby consist of 1100 mm x 1100 mm (42 in x 42 in) composite members utilizing built-up steel cruciform shapes embedded within the concrete columns.

11.3.1.4 Substructure

The vertical elements of the superstructure continue down through the substructure to the foundation. The shear walls are 840 mm (33 in) thick with a compressive strength of 55 Mpa (8,000 psi). The columns are typically 900 mm x 1200 mm (36 in x 48 in) and consist of 55 Mpa (8,000 psi) concrete. Where the columns are adjacent to the perimeter basement walls, the columns are designed to be pilasters. The gravity system in the substructure consists of a two-way flat slab system with a compressive strength of 34 MPa (5,000 psi).

11.3.1.5 Foundations

The foundation system consists of a 3 m (10 ft) thick conventional reinforced concrete mat foundation. A perimeter reinforced concrete foundation wall system consists of conventional 400 mm to 560 mm (16 in to 22 in) thick cast-in-place concrete walls.

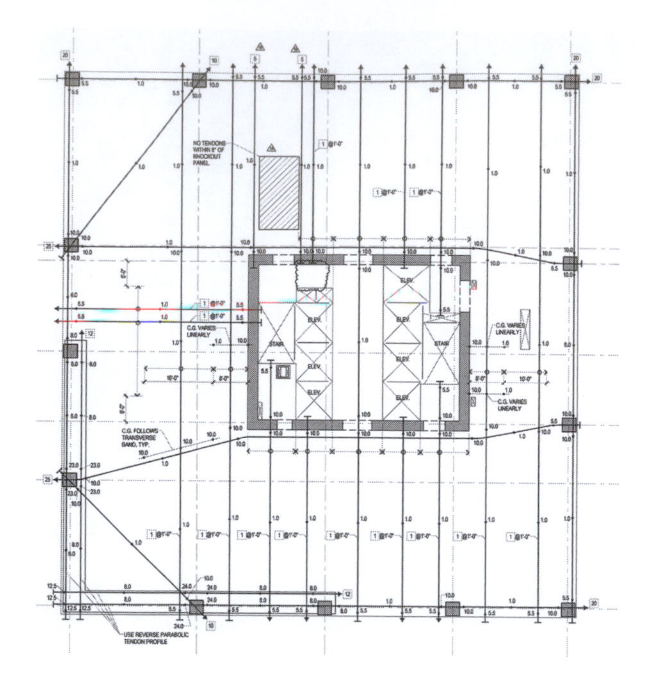

Typical Structural Floor Framing Plan, 350 Mission Street, San Francisco, CA

11.3.2 Analysis and Design Methodology

The tower lateral system did not meet the prescriptive code limits and design procedures set forth by the California Building Code (CBC 2010) and the American Society of Civil Engineers (ASCE 7-05) specifically related to the building height requirement. This exception to the prescriptive code limits and design procedures require that the structural system be classified as a non-prescriptive seismic force-resisting system. The building was required to meet the seismic performance intent of the building codes and was required to utilize San Francisco Building Code Administrative Bulletin AB-083 for seismic design, "Requirements and Guidelines for the Seismic Design of New Tall Buildings using Non-Prescriptive Seismic-Design Procedures." AB-083 also required a review of the design by a Structural Peer Review Panel, which had to coincide with the requirements stated in AB-082, "Requirements and Guidelines for Structural Design Review Procedures." AB-083, "Requirements and Guidelines for the Seismic Design of New Tall Buildings using Non-Prescriptive Seismic Design Procedures" presented requirements and guidelines for seismic structural design and building permit submittals for new tall buildings in San Francisco that use non-prescriptive seismic design procedures.

To demonstrate that the building was designed to have seismic performance at least equivalent to the intent of the code-prescriptive building, AB-083 required a three-step procedure be performed with the independent, objective, and technical peer review of the Structural Design Review Panel (SDRP).

11.3.2.1 Step One: Code Level Seismic Evaluation

The code level evaluation was used to identify the exceptions being taken to the prescriptive requirements of the SFBC and to define the minimum required strength and stiffness for earthquake resistance.

The design criteria for the code level evaluation incorporated the prescriptive provisions of R, Wo, ρ, and all other values per CBC 2010 (ASCE 7-05 Chapters 11 and 12). The strength and drift base shears were taken conservatively to act simultaneously in both directions since the core walls form part of two or more intersecting seismic force-resisting systems, per ASCE 7-05 Section 12.5.4. Scaling of the strength and drift base shears conform to ASCE 7-05 Section 12.9.4. The following structural loading combinations were used, incorporating strength design (basic for concrete per 2010 CBC Section 1605.2.1 and ACI 318-08 Section 9.2.1):

1. $1.4(D + F)$
2. $1.2(D + F + T) + 1.6(L + H) + 0.5(L_r \text{ or } S \text{ or } R)$
3. $1.2D + 1.6(L_r \text{ or } S \text{ or } R) + (f_1L \text{ or } 0.8W)$
4. $1.2D + 1.6W + f_1L + 0.5(L_r \text{ or } S \text{ or } R)$
5. $1.2D + 1.0E + f_1L + 0.2S$
6. $0.9D + 1.6W + 1.6H$
7. $0.9D + 1.0E + 1.6H$

Using the seismic load effect E as defined per ASCE 7-05 12.4.2, the load combinations 5 and 7 above became:

5. $(1.2 + 0.2S_{DS})D + \rho Q_E + L + 0.2S$
7. $(0.9 - 0.2S_{DS})D + \rho Q_E + 1.6H$

Analysis Model

A 3-D finite element computer model using the ETABS software was used to perform a linear elastic modal response spectrum analysis that conforms to ASCE 7-05 Section 12.9. The ETABS model includes the three basement levels and the entire superstructure up to and including the mechanical roof level. The model did not include the mat foundation. The following modeling parameters were used:

a. Both gravity and lateral systems were included in the model.
b. The model was pinned at the bottom of the wall shell elements and fixed at the bottom of the linear column elements. The model also contained horizontal pinned restraints at the perimeter of the below grade diaphragms.

c. Each level was modeled as a rigid diaphragm.
d. Damping was assumed to be 5%.
e. Accidental torsion was included.
f. P-Delta effects were included.
g. Property modifiers were used for the lateral and gravity systems to account for cracked section properties during a seismic event. The property modifiers used were taken from ATC-72 (Section 4.2.2 and 4.3.1) and ASCE 41-06 Supplement No. 1 (Section 6.3.1.2, Table 6.5).

		Flexure	Shear	Axial
1.	Link Beams	$0.2\ E_cI_g$	G_cA_v	E_cA_g
2.	Shear Walls	$0.5\ E_cI_g$	G_cA_v	E_cA_g

Design

The results of the linear elastic analysis were used to evaluate force levels in the lateral system components. The structural demand in each element was used to initially design each member considering appropriate load combinations and ultimate strength equations per ACI 318. The ultimate shear values for shear walls and link beams were multiplied by 1.5 to closely approximate shear demands of the MCE level evaluation. The initial prescriptive code-based member designs were used as inputs to the performance-based design evaluations at both the service and MCE Levels.

11.3.2.2 Step Two: Service Level Seismic Evaluation

A service level evaluation was required by AB-083 to demonstrate acceptable seismic performance for moderate earthquakes (essentially elastic performance with some minor yielding of ductile elements permissible). This earthquake was defined as having a 43-year mean return period (50% probability of exceedance in 30 years).

The service level evaluation of the primary structural system was required to demonstrate acceptable, essentially elastic seismic performance at the service level ground motion.

Performance Objective

The building was anticipated to have some limited structural damage when it is subjected to Service Level earthquake shaking. This damage, even if not repaired, should not affect the ability of the structure to survive future maximum considered earthquake shaking.

Design Criteria

The design criteria for the service level evaluation did not incorporate the code prescriptive provisions of R, W_o, ρ, and Cd. These values were all taken to be 1.0. The service level earthquake shaking was derived from a site-specific 2.5% damped linear uniform hazard acceleration response spectra. The displacement and strength demands computed from the linear response spectrum analysis were compared directly with the acceptance criteria stated below. The following structural loading combinations were used to determine strength demands.

a. $1.0D + L_{exp} \pm 1.0E_x \pm 0.3E_y$
b. $1.0D + L_{exp} \pm 0.3E_x \pm 1.0E_y$

where:

L_{exp} is the expected live load. L_{exp} is taken as 20% of the unreduced live load (ATC-72, Section 2.1.4). Expected material properties are assumed in the design

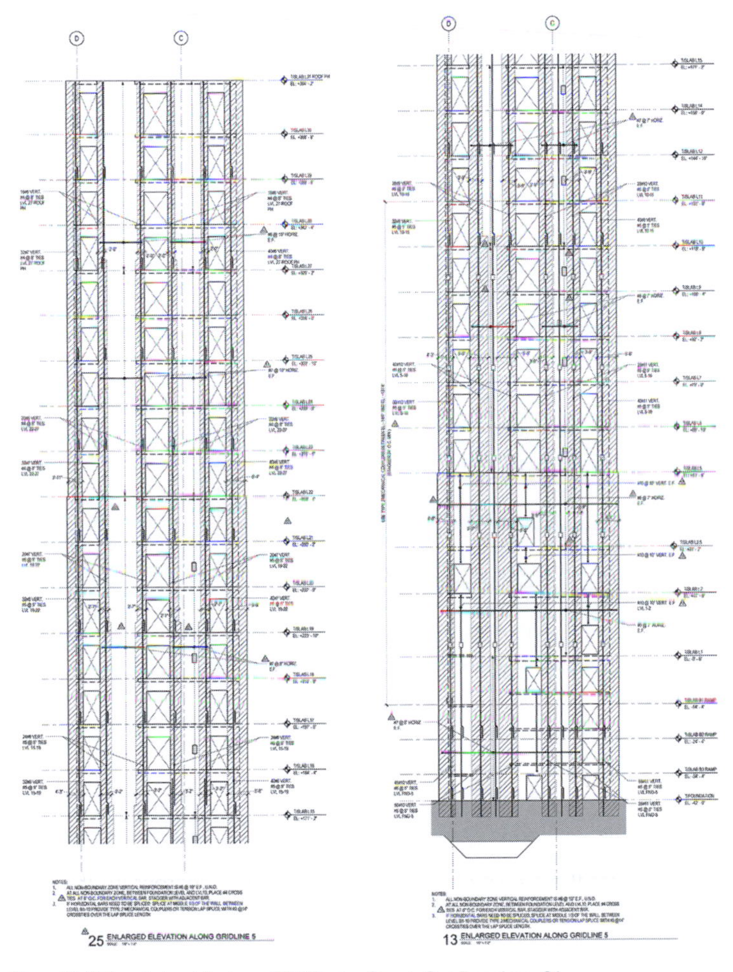

Shear Wall Core Lateral System, 350 Mission Street, San Francisco, CA

The following values were used for the service level evaluation and account for expected overstrength only. Strain hardening was excluded (PEER Guidelines, Section 7.5.2, Table 7.1).

a. Reinforcing steel: 1.17 fy
b. Concrete: 1.3 f'c

Analysis Model

A 3-D finite element computer model using the ETABS software was used to perform a linear elastic modal response spectrum analysis. The ETABS model included the three basement levels and the entire superstructure up to and including the mechanical roof level. The model did not include the mat foundation. The following modeling parameters were used.

a. Both gravity and lateral systems were modeled.
b. Bottom of the columns and walls were pinned.
c. Each level was modeled as a semi-rigid diaphragm.
d. Damping was assumed to be 2.5%.
e. Accidental torsion was not included.
f. P-Delta effects were included in the model.
g. Soil structure interaction was not included.
h. Property modifiers were used for the lateral and gravity systems to account for cracked section properties during a seismic event. The property modifiers used were taken from ASCE 41-06 Supplement No. 1 (Section 6.3.1.2, Table 6.5), ATC-72 (Section 4.2.2 and 4.3.1) and PEER Guidelines (Section 7.5.2, Table 7.2).

		Flexure	Shear	Axial
1.	Link Beams	0.3 E_cI_g	G_cA_v	E_cA_g
2.	Shear Walls	0.75 E_cI_g	G_cA_v	E_cA_g
3.	Slabs	0.5 E_cI_g	G_cA_v	E_cA_g
4.	Columns	0.5 E_cI_g	G_cA_v	E_cA_g

Note E_c was computed using $57{,}000\sqrt{f'c}$ per ACI 318-08 Section 8.5.1 with expected material strength of 1.3 f'_c. The effective shear modulus G_{eff} of walls was taken as $0.20E_c$.

Acceptance Criteria

The following acceptance criteria were evaluated:

a. The demand to capacity ratios for all lateral system components could not exceed 1.5.
b. The story drift could not exceed 0.005 of the story height in any story.

c. The shear stresses in the shear walls were generally limited to a range of $2\sqrt{f'c}$ to $3\sqrt{f/c}$.

The capacity of the lateral system components was defined as the design strength, which was taken as the nominal strength multiplied by the corresponding strength reduction factor Φ in accordance with ACI 318.

11.3.2.3 Step Three: Maximum Considered Earthquake (MCE) Level Seismic Evaluation

The MCE level evaluation was intended to verify that the structure has an acceptably low probability of collapse under severe earthquake ground motions. (This earthquake is defined as having a 2475-year mean return period—2% probability of exceedance in 50 years of exceedance or 150 percent of the median deterministic spectrum for the governing fault.) The average result, over the non-linear response history analyses, of peak story drift ratio (story drift displacement divided by the story height) was not to exceed 0.03 for any story.

Performance Objective

The building was anticipated to withstand the maximum considered earthquake shaking with low probability (on the order of 10%) of either total or partial collapse. The maximum considered earthquake level evaluation shall demonstrate, with high confidence that the structure must respond to maximum considered earthquake shaking: without loss of gravity load carrying capacity; without inelastic straining of important lateral force-resisting elements to a level that will severely degrade their strength; and without experiencing excessive permanent lateral drift or development of global structural instability.

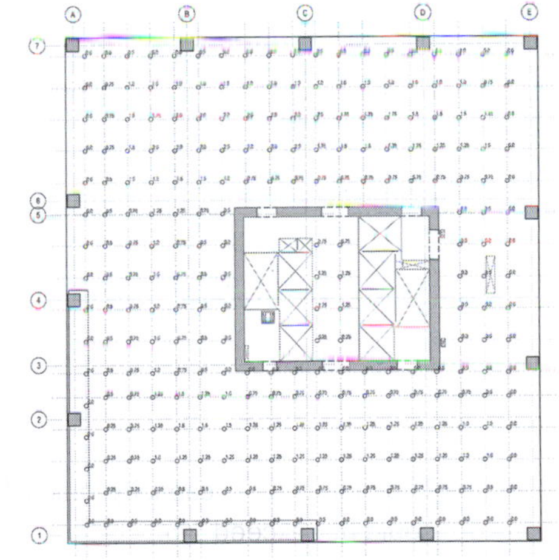

Re-Shoring System for Long-Span Flat Plate Construction Without Perimeter Frame, Engineered Slab Camber Program, San Francisco, CA

Design Criteria

The design criteria for the maximum considered earthquake level evaluation included a suite of 10 time history pairs of ground motions selected and modified using a unique amplitude scaling factor for each pair for compatibility with the target MCE shaking spectrum. The 10 time history ground motions were selected from a set of 25 ground motions based on factors such as spectral shape, peak velocity, etc. Each of the 10 time history pairs were randomly rotated and used as the seismic input in a non-linear response time history analysis. The force and deformation demands for all components and elements that form an essential part of the lateral and gravity load path were computed from the non-linear time history analyses and were compared directly with the acceptance criteria stated below.

The following structural loading combination was used to determine force and deformation demands. The demand was taken as the mean value obtained from the analyses: $1.0D + L_{exp} + 1.0E$, where L_{exp} was the expected

350 Mission Street Under Construction, San Francisco, CA

live load. L_{exp} was taken as 20% of the unreduced live load. Expected material properties were assumed in the design. The following values were used for the MCE level evaluation and account for overstrength only. Strain hardening is excluded (PEER TBI Guidelines, Section 7.5.2, Table 7.1).

a. Reinforcing Steel: 1.17 f_y
b. Concrete: 1.30 f'_c

Analysis Model

A 3-D computer model using the PERFORM-3D software was used to perform non-linear response history analyses. The PERFORM-3D model included the three basement levels and the entire superstructure up to and including the mechanical roof level. The following general modeling parameters were used.

a. Both gravity and lateral systems were present in the model. The gravity systems' contribution to the lateral stiffness was included as an equivalent frame.
b. The model was supported by pin supports at the bottom of the shear wall and basement wall meshing nodes and bottom nodes of columns
c. Viscous damping was assumed as 3.0% modal damping and 0.5% Rayleigh damping at period ratios of 0.5 and 1.5 relative to fundamental period.
d. Accidental torsion was not included.
e. P-Delta effects were included in the model.
f. The target non-linear hinge zone was between Level 5 and Level 10 (due to large typical openings and core wall section reduction).
g. Each pair of the ground motions was applied simultaneously in a bidirectional fashion. The angle of application of each ground motion pair was input at a randomly selected angle determined by the SDRP.

CHAPTER 12

ENVIRONMENT

12.1 ENVIRONMENTAL CHANGE

The average temperature rise on the planet between the start and end of the 20th century has been measured to be almost 1.5°F. The Intergovernmental Panel on Climate Change (IPCC) has concluded that most of the observed temperature increase is due to increased amounts of greenhouse gases with carbon dioxide being the major contributor. The most significant contributors to greenhouse gases resulted from human activities including the burning of fossil fuels and deforestations. The industrial revolution in the United States and Europe in the early 1900s may have been the most important period of significance. The industrial developments in China and India may take their place as the most significant contributor in the 21st century. Another 2°F temperature rise is expected to occur in this century (IPCC, 2007).

Global temperature increases will result in sea level rise, will change patterns of precipitation, and perhaps expand subtropical deserts. The arctic region is expected to see the greatest effects with the retraction of glaciers, reduction of permafrost and sea ice resulting in extreme weather patterns, extinction of species, and changes in agricultural yields. Water level rises will have catastrophic effects on developments and ecosystems that currently inhabit shorelines and waterways that interconnect with oceans.

Buildings account for 39% of the carbon emissions in the United States—more than either the transportation (33%) or industrial (29%) sectors. Most of the emissions are associated with the combustion of fossil fuels. The carbon emissions associated with manufacturing and transporting building construction materials as well as demolition or repair have even a greater impact (USGBC, 2004).

FACING PAGE
Post-Consumer Waste Plastics

12.2 ENVIRONMENTAL INSPIRATION

12.2.1 Emergence Theory

Structures that respect the growth patterns in nature will ultimately lead to minimal material use and effect on the environment. In many cases these forms are complex and require interpretation. Advanced computational and drawing tools have contributed to significant progress in developing these complex concepts. These ideas are further defined by least energy principles related to strain energy and emergence theory concepts.

Intrinsic rules and relationships shape how elementary components can comprise complex organisms and systems. These rules and relationships often orchestrate growth of the higher-level system without global oversight or guidance. Scientists have observed that these systems are organized, stable, and complex. Emergence is a theme observed in nature which suggests that complex, organized, and stable organisms and systems arise from relatively simple sub-components and their interactions without external guidance.

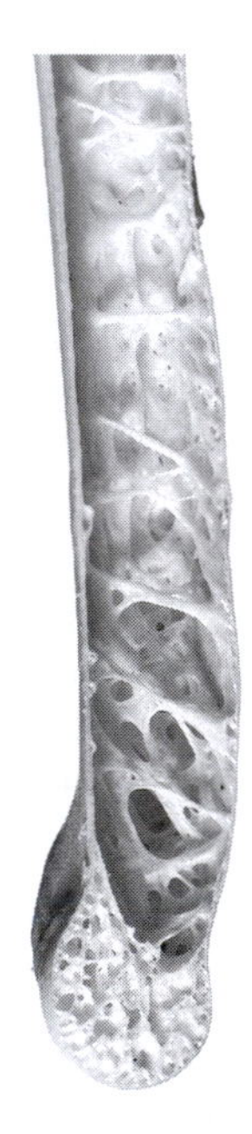

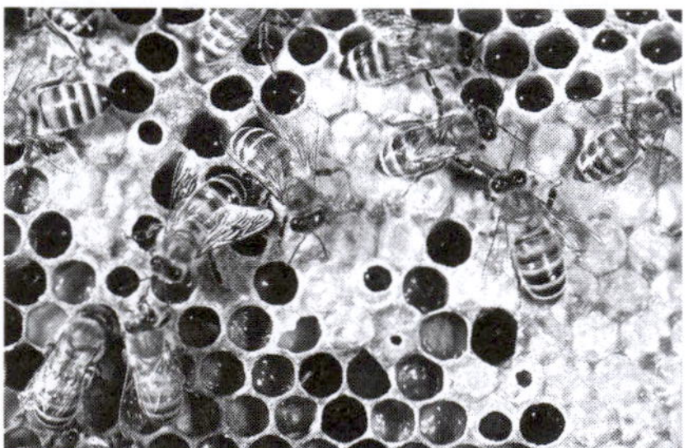

CLOCKWISE FROM TOP LEFT
Termite Colony

Bird Wing Skeleton Section

Honeycomb

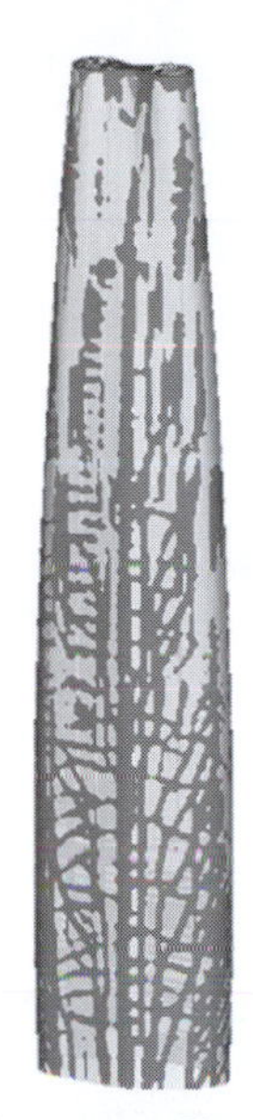

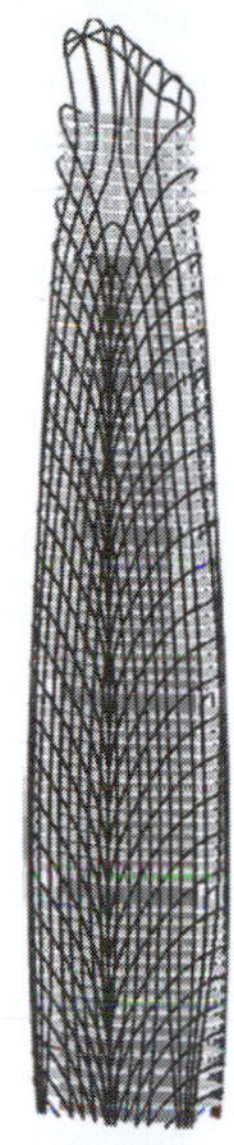

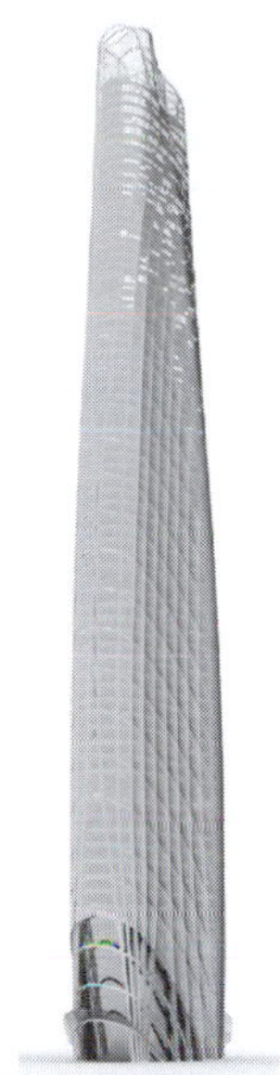

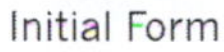

Initial Form | Emergence Theory Diagram | Final Structural Frame | Final Architectural Elevation

Final Perimeter Structural Frame, Gemdale Plaza, Shenzhen, China

An example of an emergent system is the termite colony. Although the queen termite is the only member creating children, she does not direct the termite colony as a whole. The success of the colony is entirely dependent on basic relationships between termites without supervising guidance, yet they are known to be highly organized. Also, the structures termites build are the result of an emergent process as they are built based on basic rules and regulations, but without a master plan. In spite of this, termites can build very tall and stable structures with self-cooling characteristics.

Another example of an emergent system is the bone structure of a bird's wing. Bird bones are to be light and structurally efficient to improve the capabilities of the host bird. Over time, bird wings have developed light-weight truss systems inside their wing bones. Bone framework is the result of small, unguided changes over time, and therefore is an example of emergence. A third example of emergence is the honeycomb. Bees build individual hexagonal compartments based on instinct and a highly efficient structural and storage system emerges.

The overall premise of the concept is that the collaboration of a collection of individual elements has far greater strength than if individual elements act alone.

These same principles can be applied to structures of various forms and boundary conditions. Using the emergence theory and energy principles the resulting structure represents least material and least carbon footprint.

Emergence theory and the use of optimization through strain energy principles led to a perimeter structural steel frame 25% less structural steel when compared to a conventional, rectilinear steel frame for the Gemdale Plaza structural system. This reduction in structural steel along with the introduction of enhanced seismic systems such as the Pin-Fuse Frame™ resulted in a 30% reduction of carbon for this 71 story, 130,000 square meter (1.4 million square feet), 350 m (1150 ft) tall tower in Shenzhen.

12.2.2 Fibonacci Sequence

Some would argue that the mathematics used to define Fibonacci Sequence may be the numerical definition of life. Galaxies, hurricanes, hair growth, plant growth, and the proportions of the human body are all based on the same sequential principle. These natural forms, based on binary numbers, 0 and 1, represent the most efficient structures.

In 1937 when MIT student Claude Shannon recognized that the *and*, *or*, or *not* switching logic of Boolean algebra developed almost one hundred years earlier was similar to electrical circuits. This yes–no, on–off approach to defining operations described in Shannon's thesis led to the practical use of binary code in computing, electrical circuits, and other applications. This was the breakthrough to Gottfried Wilhelm Leibniz's 17th-century search for finding a system of converting logic verbal statements to mathematical ones. The use of ones and zeros can be assembled into binary strings where 8 digits can be assembled to represent 256 possible combinations of different letters, symbols, or instructions.

The binary digits found in these logic formations are prevalent in many other numerically defined forms, particularly those seen in nature. Those natural forms highlight organization, efficiency, and proportion. The golden spiral, mathematically defined by the Fibonacci Sequence, is rooted in logic formations created by binary numbers. The Fibonacci Sequence is a

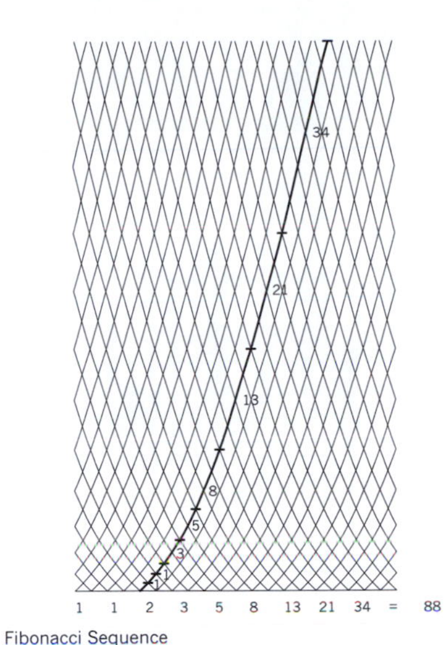

Fibonacci Sequence

Frame Geometry Definition based on Fibonacci Sequence, Wuxi Times Square, Wuxi, China

Ø=1.618

1.0

1/Ø =0.618

13

8

2 1 1

3

5

Fibonacci Sequence/Golden Spiral

series of numbers where the subsequent number in the sequence is equal to the sum of the previous two numbers starting with 0 and 1. The sequence of numbers is 0, 1, 1, 2, 3, 5, 8, 13, 21, 34, 55, 89, 144, 233, 377 ... More interesting is the multiplicative relationship between any two numbers as each number approximates the previous number by the golden section multiple. This numerical ratio or proportional constant (for instance, $\varnothing = \frac{13}{8} = 1.618$) converges on the divine limit as the numbers in the sequence grow. The value of this constant converges on 1.618.

The mathematical definition of the sequence is as follows:

$$F_n = F_{n-1} + F_{n-2}$$

where:

F = Fibonacci number
n = number in the sequence

and with the initial numbers in the sequence defined with binary digits of 0 and 1:

$$F_0 = 0$$
$$F_1 = 1$$

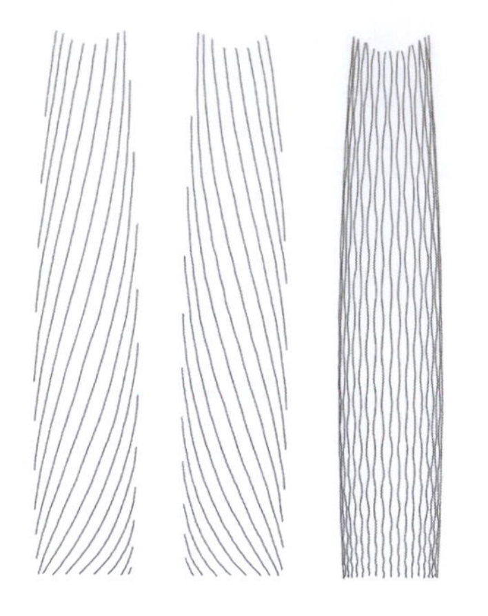

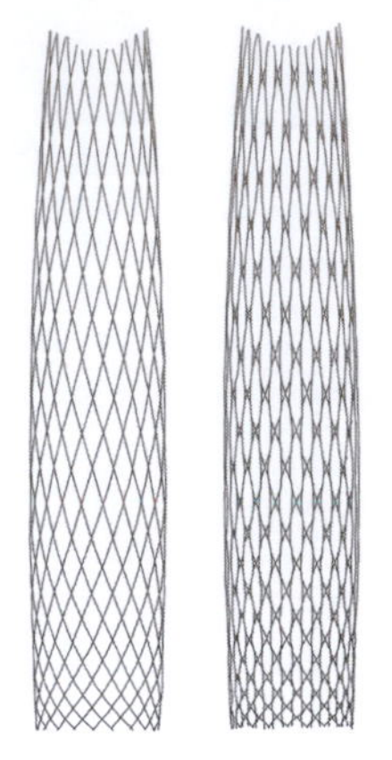

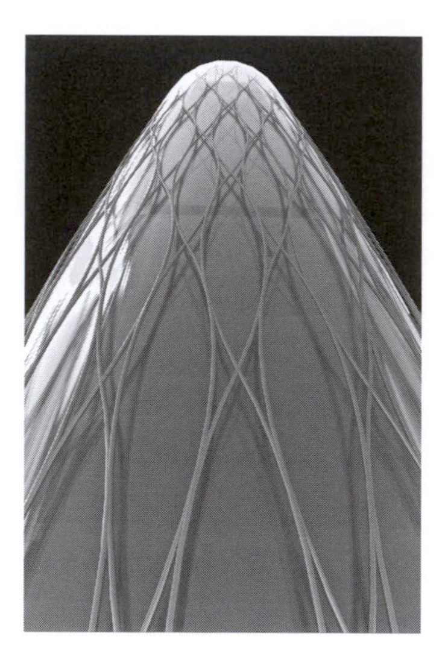

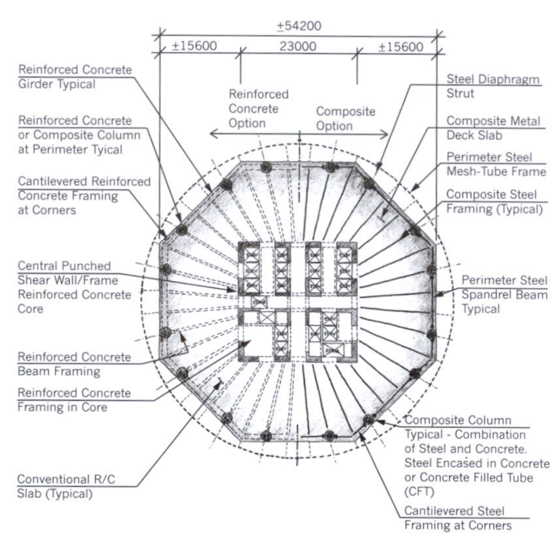

CLOCKWISE FROM TOP LEFT

Elevation Drawings of Evolution of Structural System

Fibonacci Sequence Inspiration, Wuxi Times Square, Wuxi, China

Structural System Plan, Wuxi Times Square, Wuxi, China

Exterior Frame, Wuxi Times Square, Wuxi, China

Exterior Frame, Wuxi Times Square, Wuxi, China

Wuxi Times Square, Wuxi, China

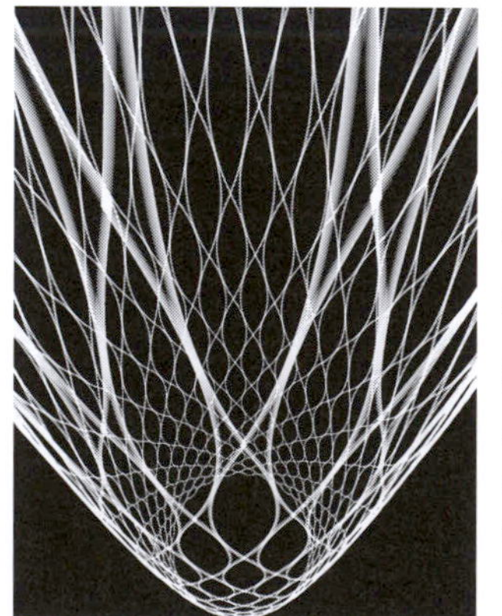

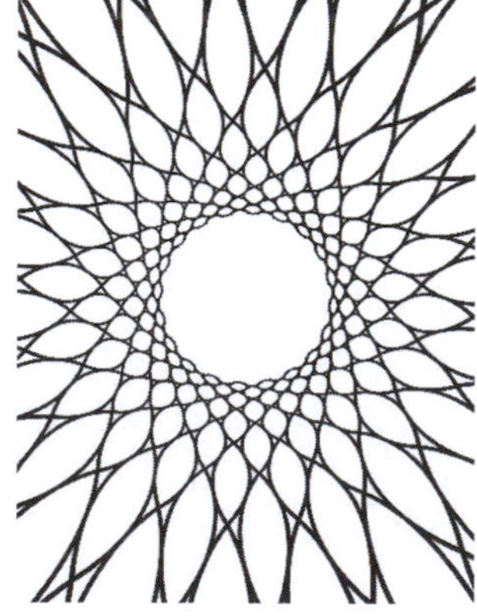

So, for the first number in the sequence ($n = 1$):

$$F_1 = F_{(1-1)} = F_0 = 0$$

For the second number ($n = 2$):

$$F_2 = F_{(2-1)} + F_{(2-2)} = F_1 + F_0 = 0 + 1 = 1$$

For the third number ($n = 3$):

$$F_3 = F_{(3-1)} + F_{(3-2)} = F_2 + F_1 = 1 + 0 = 1$$

For the fourth number ($n = 4$):

$$F_4 = F_{(4-1)} + F_{(4-2)} = F_3 + F_2 = 1 + 1 = 2$$

For the fifth number ($n = 5$):

$$F_5 = F_{(5-1)} + F_{(5-2)} = F_4 + F_3 = 2 + 1 = 3$$

For the sixth number ($n = 6$):

$$F_6 = F_{(6-1)} + F_{(6-2)} = F_5 + F_4 = 3 + 2 = 5$$

For the seventh number ($n = 7$):

$$F_7 = F_{(7-1)} + F_{(7-2)} = F_6 + F_5 = 5 + 3 = 8$$

And so on.

Despite a seemingly infinite variety and diversity of plant growth, nature employs only three fundamental ways of arranging leaves on a stem. The first is distichous like corn, the second is decussate like mint, and the third is spiral phyllotaxis which represents 80% of the higher order plants where the rotation angle between leaves is the golden angle of 137.5° (360°/Ø²). With this spiral growth, no future leaf overshadows a predecessor allowing each leaf to receive maximum sunlight and rain—all based on the binary numbers of 1 and 0.

The gravity load-resisting system consists of conventional reinforced concrete, structural steel, or a combination of the two. The exterior frame, designed to resist lateral loads only, is connected through floor diaphragms to the interior gravity frame. The geometry of the perimeter frame is defined by the Fibonacci Sequence, with bracing members more vertical with greater spacing at the top of the structure, where cumulative lateral forces requiring resistance are smaller, and more horizontal with tighter spacing near the base, where cumulative lateral forces are largest and so require greatest resistance.

12.2.3 Genetic Algorithms

The Al Sharq Tower is to be located in Dubai, United Arab Emirates. The plan form of the structure is based on nine adjoining cylinders. Since traditional perimeter columns are not desired, a cable-supported perimeter is implemented. Early efforts by the architectural and structural design teams to generate an aesthetically appealing profile followed classic geometric definitions such as that of a helix.

The 102 story residential tower has a 39 m x 39 m (128 ft x 128 ft) floor plan, a height of 365m, and therefore an aspect ratio nearing 10:1. The proposed structural system consists of reinforced concrete systems with

Al Sharq, Dubai, UAE

STRUCTURAL STEEL
CORE SHEAR WALL FORMWORK
SLAB FORMWORK
SLAB SHORING
ADDITIONAL SHORING PER SOM RF CONCRETE SPEC.
RE-SHORING AT THE SLAB EDGE
TO SUPPORT PERIMETER WITHOUT CABLES

Al Sharq Construction Sequence, Dubai, UAE

Darwinian Evolution

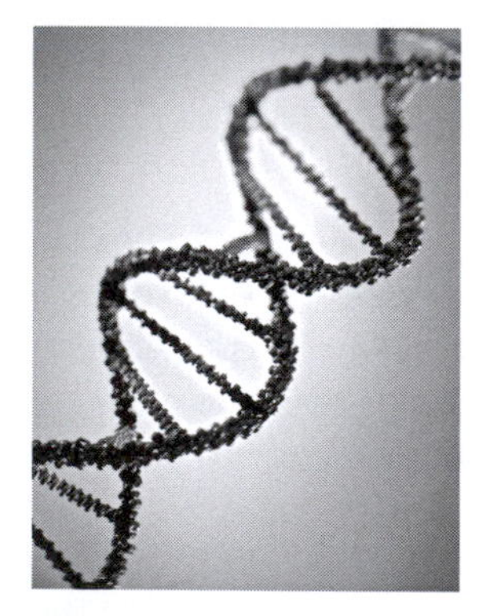

DNA Strand

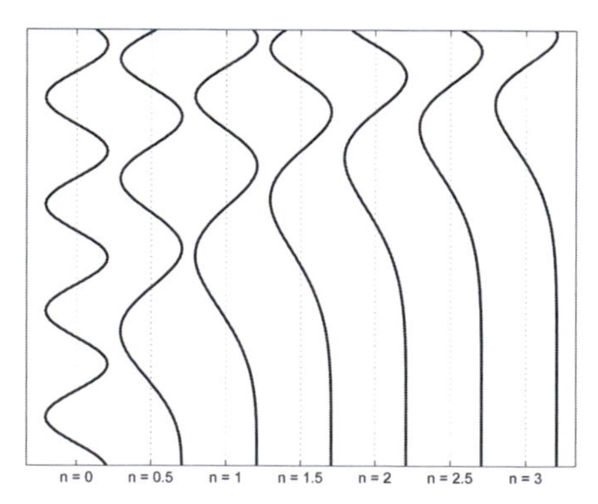

Various Tapered Helices, n=2,
Parabolic Helix Used for Al Sharq Tower

perimeter spiraling high-strength galvanized steel cables. The lateral system is composed of intersecting sets of parallel shear walls and perimeter high-strength galvanized steel cables. The perimeter cable system consists of approximately 70 kilometers (44 miles) of high-strength galvanized steel cables. Initial cable profiles suggested a helical formulation for each. The initial helix definition is propagated to each perimeter cylinder.

12.2.3.1 Cable Profile Influenced by Genetics

Observation of principal stress as a means of form generation is well established in nature. This is demonstrated by the growth patterns of nautilus shells, fiber-reinforcement of palm tree branches and structure of bones which mimic the flow of applied forces. Principal stress observation is undertaken for the preliminary identification of the optimal cable profile for the Al Sharq Tower.

An investigation of principal stresses over the building's perimeter skin is conducted to determine how the form of the building might react to lateral loads. Analysis results show that for the bundled cylinder plan the corner modules exhibit vertical tension (or compression) at the base and transition to 45° (shear) near the top.

Since cables are tension-only members, their most efficient orientation is in alignment with the direction of principal stress. Thus, a cable profile based on observed principal stress trajectories is needed. To define transition of principal stress trajectories, a modified helical formulation is employed. The modified-helical formation is shown in Equations 1 through 3, where z_{Total} is the total height of the building, and n is an adjustable parameter that defines the rate of pitch transition over the height of the structure. The ratio of current to total height raised to the power n alters the cable pitch as a function of tower height. A value of $n = 0$ yields no transition of pitch, $n = 1$, yields a linear transition from vertical to 45° over the height of the structure, $n = 2$ yields a parabolic transition, etc. From observation of the principal stresses, it is determined that the transition of principal stress orientation over the face of the windward corner modules is:

$$X(z) = r\cos(t) \tag{1}$$

$$Y(z) r\sin(t) \tag{2}$$

$$t = z\left(\frac{z}{z_{Total}}\right)^{n} \tag{3}$$

approximately parabolic ($n = 2$). Thus, a parabolic-tapered helix definition could be used to fully define the cable profile at each perimeter cylinder over the height of the structure. As can be observed, the illustrated parabolic-taper helix definition closely matches that of the corner module of the windward face of the principal stress contours.

Observation of principal stresses at the building perimeter for the determination of cable profile is reasonable if the exterior were monolithic and homogeneous. The perimeter is actually a series of discrete tension-only cables. Thus, the principal stress investigation may provide a rational basis of global-perimeter load paths, but further investigation is needed to determine an optimal cable profile.

12.2.3.2 Cable Profile Optimization Using Genetic Algorithms

With a rational basis of perimeter load path established, further investigation is sought to develop an efficient cable profile for the resistance of lateral

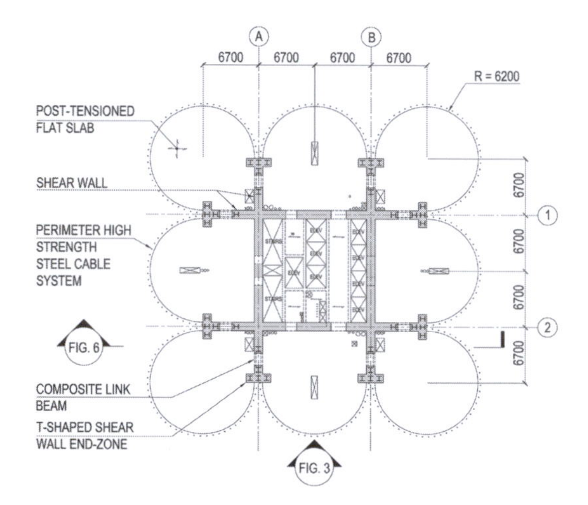

Al Sharq Floor Framing

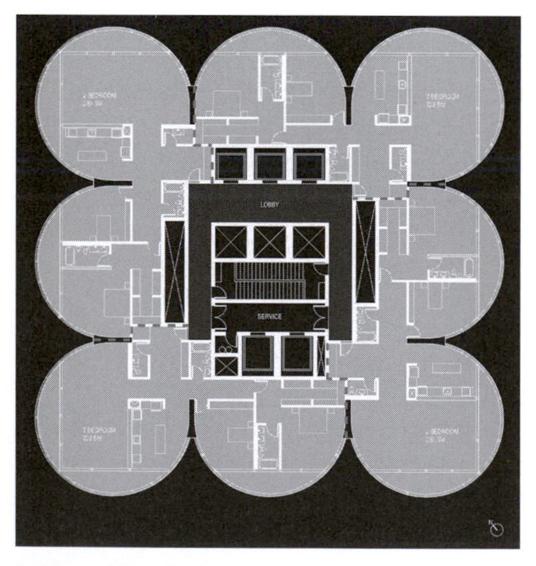

Architectural Floor Plan, Al Sharq Tower, Dubai, UAE

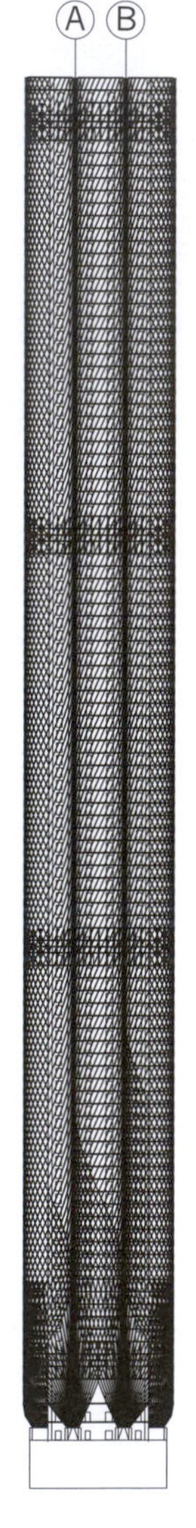

Elevation, Al Sharq Tower, Dubai, UAE

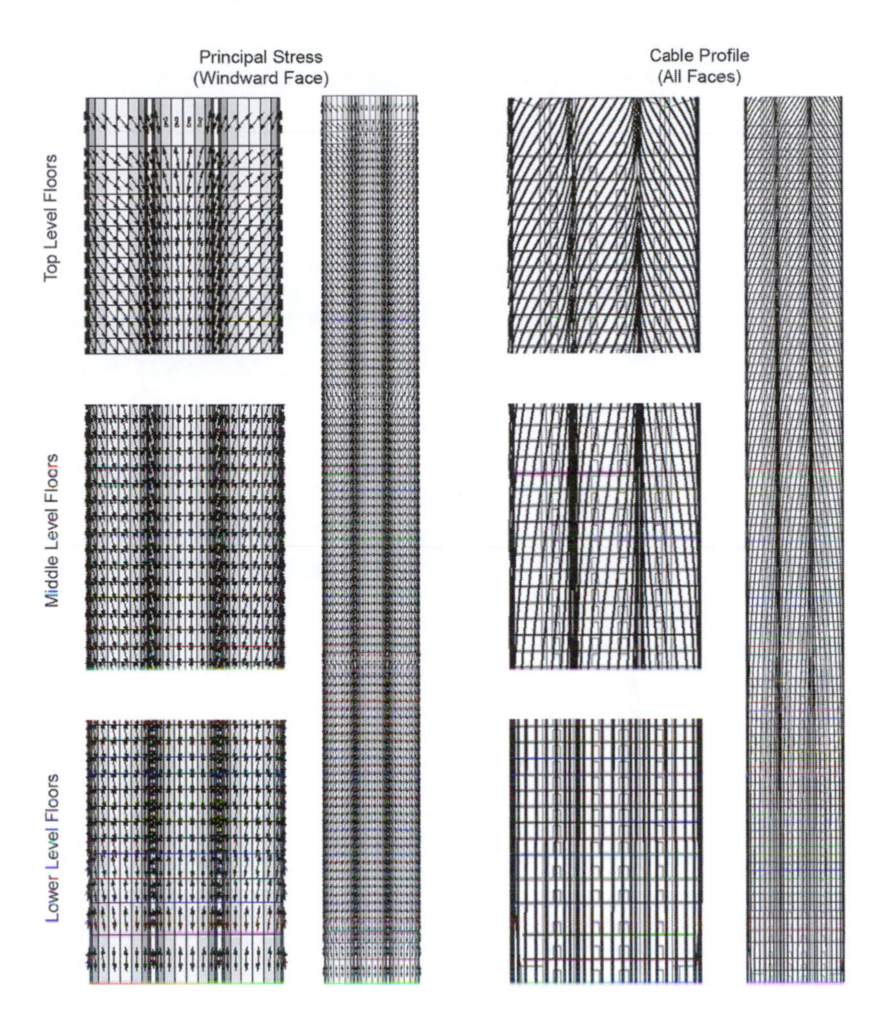

Principal Stress Analysis and Cable Profile

loads. Improved performance is pursued through the employment of a genetic algorithm (GA) optimization routine. GA optimization considers a pitch which varies over the height of the tower. In what follows, a general description of the employed GA is provided.

Genetic algorithms have been used in a wide range of applications for improved performance in numerous trades such as the aerospace, automobile, and medical industries. This simple, yet robust algorithm facilitates multi-variable and multi-objective searches in large, often poorly defined, search spaces. Early investigations of evolutionary algorithms were conducted by Holland (1975) and inspired from observations made by Darwin (1859). GA is a heuristic optimization method which utilizes trial and error of mass populations as a basis of optimization. To demonstrate GA concepts, a simple truss optimization problem is illustrated in the following text.

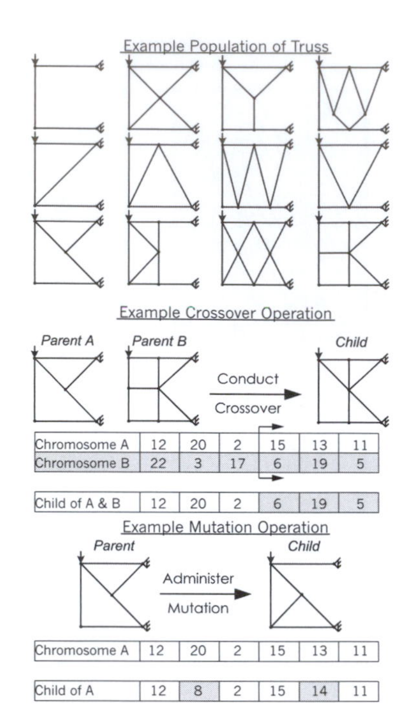

Genetic Algorithm Illustration

View of Apartment, Al Sharq Tower

For GA optimization to begin, an initial population must first be generated. A population is a group of candidate-solutions. For the example truss problem, a population would consist of a set of potential truss configurations. Each truss would have a different member configuration but the loading and boundary conditions would be the same.

With an initial population generated, candidate-solutions are evaluated. Their fitness, or score, is determined by a fitness function. For this example a truss's fitness is the sum of normalized deflection and normalized weight. This GA is a minimization algorithm, thus the sum of the normalized values is taken for the fitness. Both deflection and weight must each be normalized to minimize bias in the fitness score. Analysis software can be used to quickly determine the deflection and weight of each truss in the population. Increased weight and deflection increase the fitness of a candidate-truss and therefore diminish its chances of being selected by the GA for inclusion in future generations.

The initial population is the first parent population and is used to generate the child population. The child population is a new set of candidate-solutions which are derived from the parent population. The child population is to be the same size as the parent population. Each member of the child population is to be generated using GA operators. Parameters to be optimized by the GA are contained in a vector of values termed a chromosome.

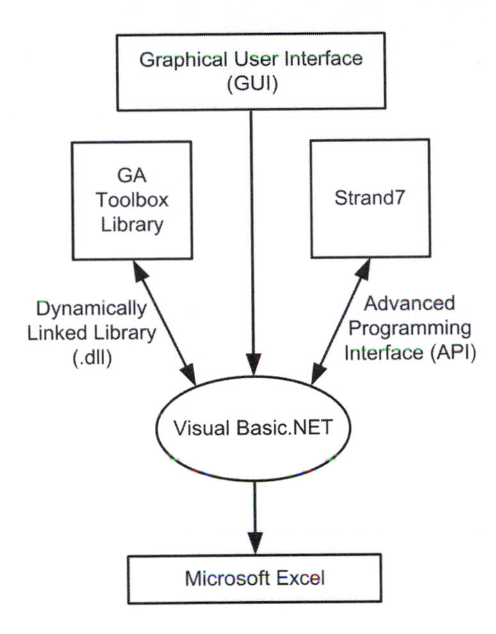

Implementation of Genetic Algorithm

The first type of GA operation is called "crossover." A crossover operation takes two parents and combines characteristics from each parent to form a child. The second type of GA operation is called "mutation." A mutation operation takes one parent and alters one or more characteristics of the parent to form a child.

After the child population is generated, each child is evaluated and fitness determined. Next, parent and child populations are combined into a single pool of candidate-solutions. The pooled set is ranked according to each member's fitness score. For the truss example problem, the truss with the lowest fitness score is considered best and the truss with highest fitness score is considered worst. With the pooled parent and child populations ranked, the top 50% are elected to be the parent population for the next generation. The remaining trusses are discarded.

To implement GA for the optimization of Al Sharq cable filigree, several tools are needed. Visual Basic.NET is a general purpose programming environment well suited for conducting GA operations, interaction with finite element software, and collection of results. Finite element analysis software Strand7 is utilized for analysis of GA-generated cable profiles.

With the concepts of the genetic algorithm described, its application to the optimization of cable filigree of the Al Sharq Tower is now considered. GA operations are to optimize cable pitch at each floor. As already observed in the principal stress analysis, optimal cable pitch may vary over the height

$$J_1 = \frac{1}{(Cable\ Area * Total\ Cable\ Length * Roof\ Drift)} \quad (7)$$

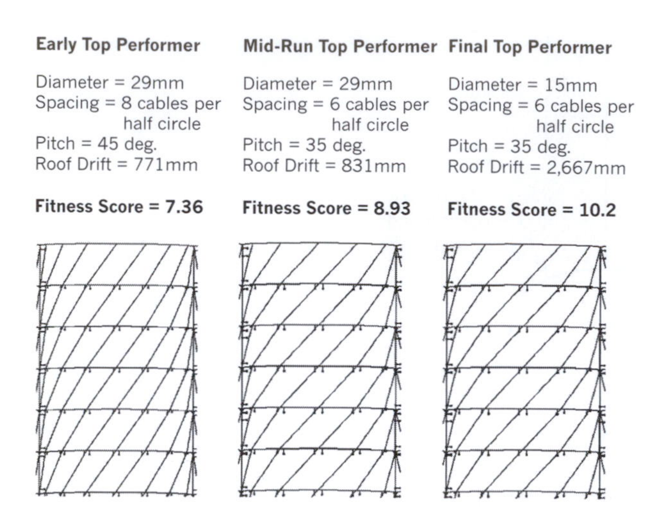

Summary of Fitness Score Results,
Al Sharq Tower, Dubai, UAE

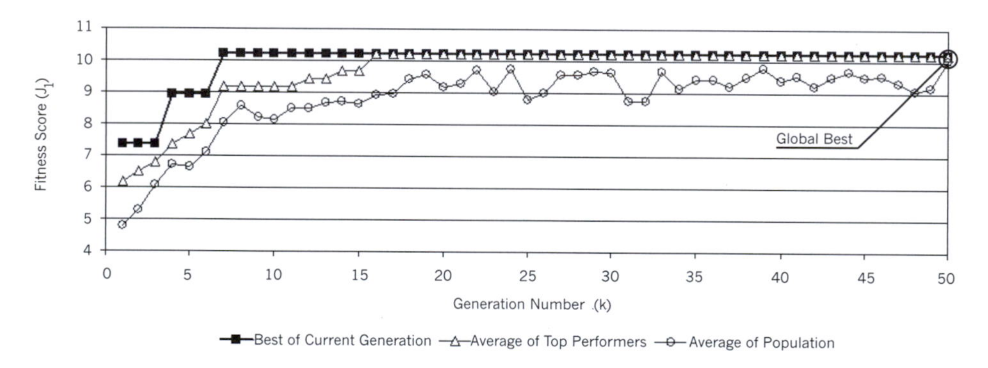

Results of Genetic Algorithm Optimization,
Al Sharq Tower, Dubai, UAE

of the tower. With this in mind, GA optimization allows pitch at each floor to be varied. Thus, 102 variables are concurrently optimized. Fitness function is the normalized roof drift.

A total of 500 generation-cycles are conducted with a population size of 10, thus evaluating a total of 5000 potential cable filigree configurations. Fitness scores steadily improve until approximately generation 350. The top performing solution from generation 500 reveals a cable profile which is very similar to the parabolic profile determined in the previously discussed principal stress-based cable profile study.

Theoretical Parabolic Tapered Helix Profile,
Al Sharq Tower, Dubai, UAE

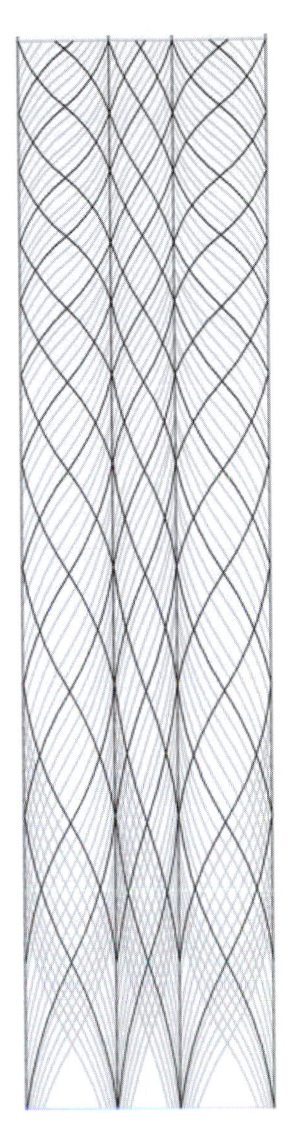

Final Interpreted Cable Profile,
Al Sharq Tower, Dubai, UAE

Early efforts to determine an efficient and rational cable profile in response to lateral loads have yielded a cable profile derived from the observation of principal stress contours and confirmed by GA optimization. The optimal cable profile follows a parabolic-helical definition (n = 2) which closely matches the principal stress contours observed; vertical at the base transitioning to 45° at the top.

12.3 SOLUTIONS FOR A CHANGING ENVIRONMENT

12.3.1 Automobile Analogy

When people consider purchasing an automobile, one of the primary considerations is gas mileage—the greater the gas mileage, the fewer natural resources expended and the smaller the carbon footprint. However, how many consider where the car is made, what materials were used to make it, if the car is safe, whether it requires frequent maintenance, and what the vehicle's expected life is? All of these considerations contribute greatly to the vehicle's environmental impact and in some cases could be the most significant impact when compared to its operation. Now ask the same questions of building a new structure (or even retrofitting an existing structure):

- Where do the materials for the structure come from, including the natural resources, fabrication, and shipment to the project site?
- What is the anticipated construction time?
- The structural system—will it behave naturally in an earthquake?
- The performance of the structural system—will the structure require maintenance during its service life? Will it require repair or replacement after a significant seismic event?
- The expected service life—when will the structure require replacement?

The carbon emissions associated with building construction range from 10 to 20% of the total carbon produced to operate it for a 50-year building life. As future buildings are designed to net-zero energy standards, the carbon associated with the initial construction will represent 100% of the total carbon emitted. In addition, it typically takes 20 years for the carbon associated with a typical building operation to outweigh the initial carbon required to build the structure.

12.3.2 Conventional versus Enhanced Seismic Systems

The engineering community has made significant advancements for the design of structures in regions of high seismic risk but most of these developments have focused on life safety, with modest focus on performance or long-term economic viability. Essentially no attention has been paid to the environmental impact of these structures. Structures that naturally coexist with their site conditions produce the most efficient designs and the most cost-effective long-term solutions, and have the least impact on the environment.

Imagine structures that behave elastically with least required materials even when subjected to the most extreme seismic events. Imagine

Friction Pendulum Base Isolator
(courtesy of Earthquake Protection Systems)

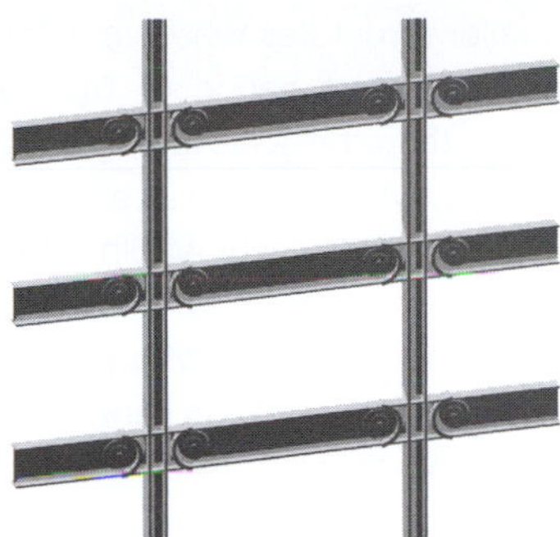

Frame with Pin-Fuse Joints®
(Patent US No. 7,000,304 & No. 6,681,538)

these structures are designed on the basis of natural behavior principles rather than conventional approaches to design. Improvements were made to steel beam–column moment connections following the non-ductile performance of many frames in the 1994 Northridge Earthquake. Other systems, such as the RBS (Reduced Beam Section or Dogbone) connection and the Slotted Web Connection, were developed. However, these systems, and others, plastically deform and incur damage during major earthquakes. We must develop systems that dissipate energy, deform elastically instead of plastically, and allow the building to be placed immediately back into service after the earthquake.

Allowing controlled movement with the dissipation of energy in structures during earthquakes is important. This movement can occur at its base or within the superstructure itself. Seismic isolation is an excellent solution to decoupling structures from strong ground motions for low- to mid-rise structures but does become somewhat more difficult to implement

350 Mission Street, San Francisco, CA

350 Mission Street Under Construction, San Francisco, CA

in taller structures where gravity loads are very large, uplift conditions could exist, and the period of the superstructure is typically long—sometimes longer than the achievable period of an isolation system.

When structures are fixed to their foundations, this movement must be designed to occur within the joints of the superstructure. Pin-Fuse seismic systems are designed to maintain joint fixity throughout the typical service life of the structure. When a significant seismic event occurs, forces within the frame cause slip in joints through friction-type connections. This slippage alters the characteristics of the structure, lengthening the period, reducing the forces attracted from the ground, and provides energy dissipation without permanent deformation.

12.3.3 Reduction of Seismic Mass

The most efficient and environmentally responsible structures are those with least mass and incorporate structural solutions informed by nature. Reducing seismic mass can be accomplished with the use of light-weight materials such as light-weight concrete that has 25% less mass than normal-weight concrete. However, other concepts can be introduced that further reduce mass. In all concrete structures there are areas that include significant amounts of concrete only because of conventional construction practices. For instance, the concrete needed in the middle–middle strip of a two-way reinforced concrete floor framing system could be reduced by 50% or more by introducing more scientific systems. If concrete in these areas could be displaced where it is not required, this reduction could be achieved. An inclusion system, perhaps one that includes post-consumer waste products could be used. The patented Sustainable Form Inclusion System™ (SFIS) accomplishes this both by creating voids within the concrete framing system and by using materials such as plastic water bottles, plastic bags, waste Styrofoam, etc. that would otherwise be placed in landfills.

Post-Consumer Plastics Formed Into SFIS™ Rectangular Units

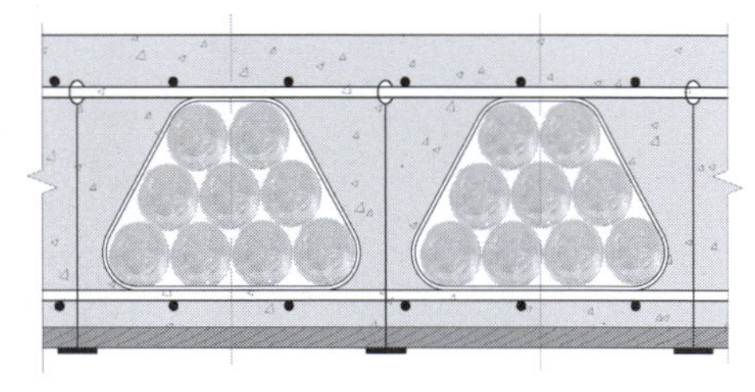

Water Bottle Concept, SFIS™

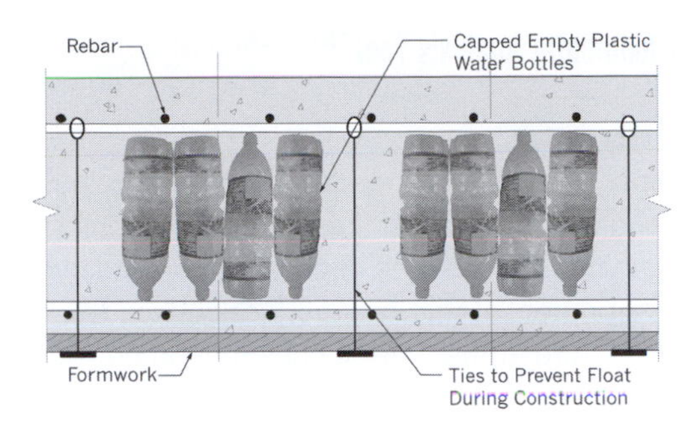

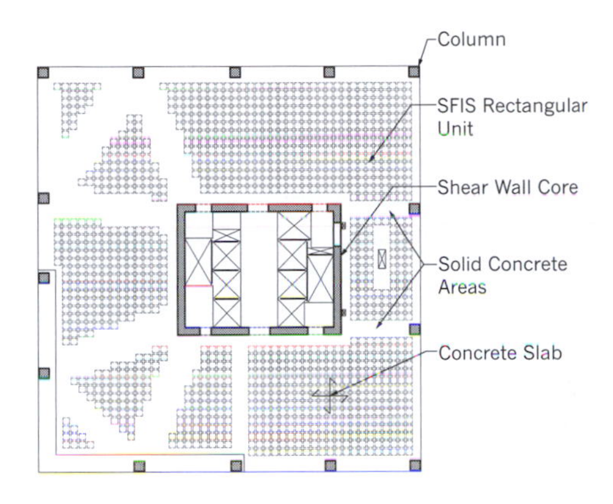

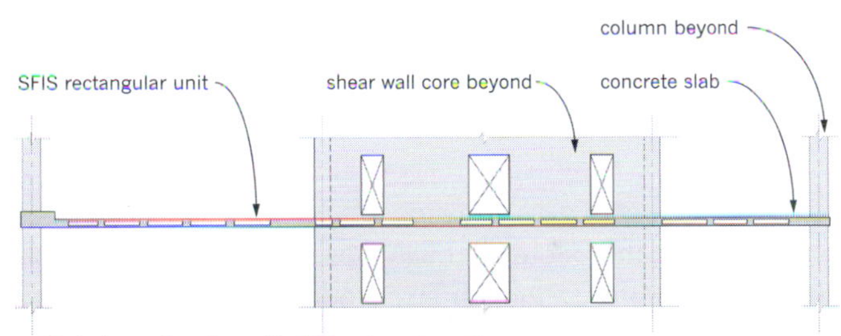

Typical overall section with SFIS rectangular units

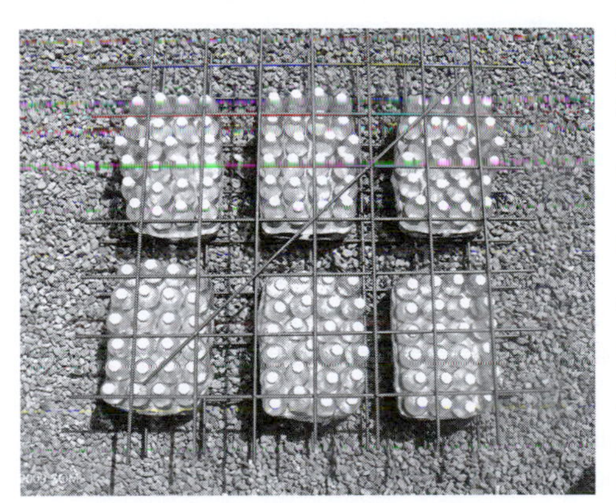

TOP TO BOTTOM

Sustainable Form Inclusion System™ (Patent US No. 8256173)

Typical Framing Plan with SFIS™ Rectangular Units in a High-Rise Structure

Typical Overall Section with SFIS™ Rectangular Units in High-Rise Structure

Typical Reinforcing Mock-Up with Water Bottles Used in SFIS™ System for High-Rise Structure

12.3.4 The Environmental Analysis Tool™

12.3.4.1 The Basis of Evaluation

Up until now, most, if not all, of the efforts made in calculating the carbon footprint have been associated with the operations of buildings, with little or no focus on the structure at the time of construction and over its service life. The patented Environmental Analysis Tool™ calculates the expected carbon footprint of the structure at the time of construction considering its location and site conditions. Based on the structural system considered, a complex damage assessment is performed accounting for the expected seismic conditions and anticipated service life, whether the structure comprises a code-defined conventional system or enhanced structural system.

12.3.4.2 Carbon Mapping Early in Design

Environmental impacts of structures need to be considered just as much as available materials, constructability, and cost at the earliest of stages of design. It is important that the carbon footprint assessment is accurate even with a limited amount of known information. The Environmental Analysis Tool™ is capable of calculating the structure's carbon footprint even when only the following information is known:

1. The number of stories (superstructure and basement).
2. The total framed area in the structure or area average area per floor.
3. The structural system type.
4. The expected design life.
5. Site conditions related to expected wind and seismic forces.

With this limited amount of data, the program references to a comprehensive database containing the material quantities for hundreds of previously designed SOM structures, although any credible database can be used to define expected materials in the structure. Curve fitting techniques are used to consider building height relative to low, moderate, or high wind/seismic conditions. Superstructure materials include structural steel, reinforced concrete, composite (combination of steel and concrete), wood, masonry, and light metal framing. Foundation materials include reinforced concrete for spread/continuous footings, mats, and pile-supported mats.

The selected seismic-resisting system is important to the carbon footprint over the life of the structure. The contribution of carbon related to damage from a seismic event could account for 25% or more of the total carbon footprint for the structure. The basis of design includes conventional code-based systems with the option to select enhanced seismic systems such as the Pin-Fuse Seismic Systems, seismic isolation, unbounded braces, and viscous dampers. The repair required for damage associated with each system is considered. The program uses this fundamental information to

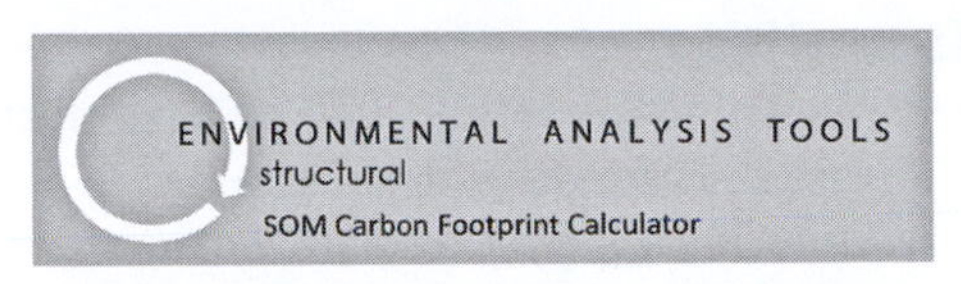

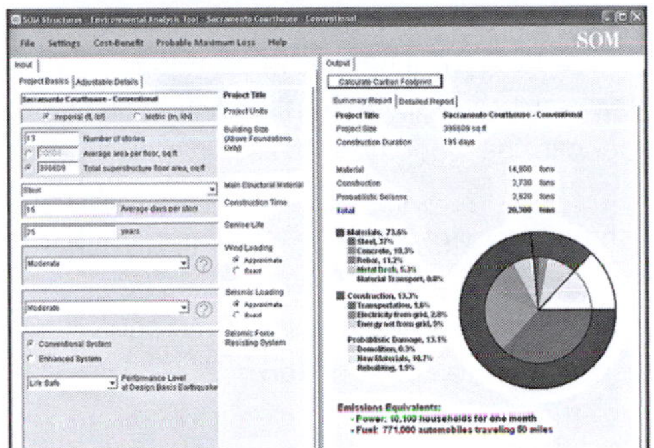

Environmental Impact Analysis

Sacramento Criminal Courthouse

Parameters:

- Steel (MF + BRBF)
- Conventional seismic system
- Steel Quantity: 16.5psf
- 396,609 sq. ft.
- 13 stories
- 25 yr life-cycle
- Seismic performance level: "Life Safety"

Estimated Carbon Footprint:

- Material = 14,900 tons CO2
- Construction = 2,730
- Seismic Damage = 2,620
  - TOTAL = 20,300 tons CO2

20,300 tons

Power: 10,100 households for one month
Fuel: 771,000 automobiles traveling 50 miles

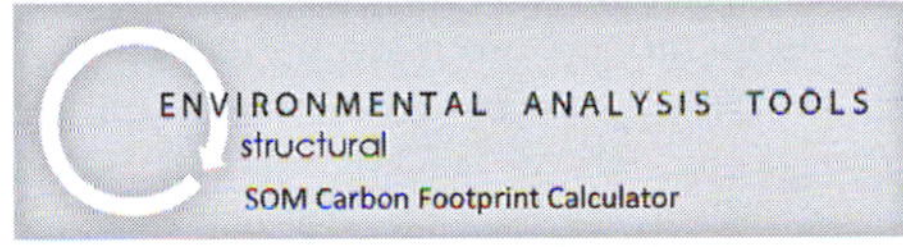

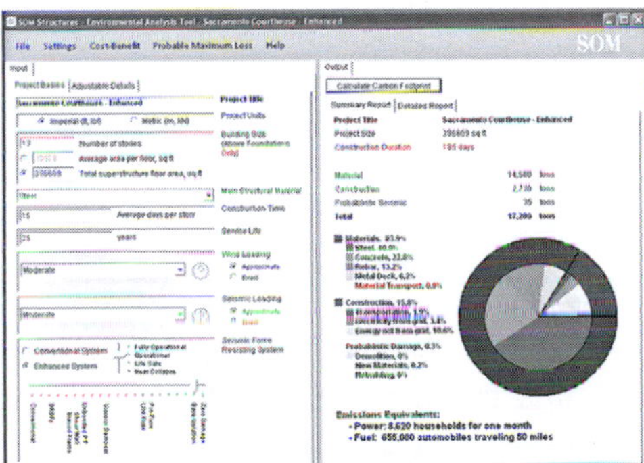

Environmental Impact Analysis

Sacramento Criminal Courthouse

Parameters:

- Steel (MF + BRBF + Base Isolation)
- Enhanced seismic system
- Steel Quantity: 15.5psf
- 396,609 sq. ft.
- 13 stories
- 25 yr life-cycle
- Seismic performance level: "Operational"

Estimated Carbon Footprint:

- Material = 14,500 tons CO2
- Construction = 2,730
- Seismic Damage = 35
  - TOTAL = 17,200 tons CO2

17,200 tons

Δ = 3,100 tons

Power: 8,620 households for one month
Fuel: 655,000 automobiles traveling 50 miles
15% reduction

(top) Carbon Footprint Calculation without Considering Seismic Isolation (Patent US No. 8256173)

(bottom) Carbon Footprint Calculation Considering Seismic Isolation (Patent US No. 8256173)

calculate the time and methods of construction, the fabrication and transportation of material, the labor required to build the structure, and laborers' transportation needs, among other things. Using this limited amount of information, an early but accurate assessment of the structure's carbon footprint can be made.

12.3.4.3 Carbon Mapping in Advanced Stages of Design

Every input variable in the Environmental Analysis Tool™ can be overridden by the user. This allows specific design information to be included in the evaluation after more comprehensive engineering is complete. For instance, specific structural steel, concrete, and rebar quantities can be used to accurately define the carbon footprint associated with materials. When specific supply locations for materials are known, the transportation distances can be used. Construction time can be modified based on anticipated complexities.

Default fragility curves that anticipate damage can be modified to accurately reflect the non-linear behavior of the structural system. Finally, the user can override specific assumptions made for carbon emission equivalents of anything from material to transportation to construction.

12.3.4.4 Environmental Analysis Tool™ Program Details

Equivalent carbon dioxide emissions associated with the structural system of a building may be categorized as those resulting from the following three major components: (1) materials used to manufacture the structure; (2) construction; and (3) probabilistic damage due to seismicity. For instance, the primary material in a steel superstructure includes structural steel, concrete, rebar and metal deck with manufacturing and transportation of the material from the source to the project site considered. Construction includes transportation of materials on the site, transportation of workers to and from the site, electricity required from the grid, and energy required from off-grid source including such items as on-site diesel-powered generation. Finally, probabilistic damage due to seismicity includes demolition of damaged areas, repair, and replacement of structural components.

The carbon calculator considers the measurement of *equivalent* carbon dioxide emissions for a building structure. "Equivalent" is used to account for other gases besides carbon dioxide (CO_2) that are considered to be greenhouse gases and contribute to the total, 100-year global warming potential (GWP) of the structure in question, which is given in units of CO_2e, or equivalent carbon dioxide. To sum up the contributions from each of these gases to the total GWP, factors are assigned to each gas based on molecular weight using carbon dioxide as the benchmark. An example is methane, its GWP is 21 in equating it to CO_2.

12.3.4.5 Cost–Benefit and PML

One of the most difficult obstacles to introducing advanced engineering components into structures is first cost. Many of these systems will require a higher initial investment; however, when considered over the life cycle of the structure, cost–benefit and Probable Maximum Loss (PML) are extremely important.

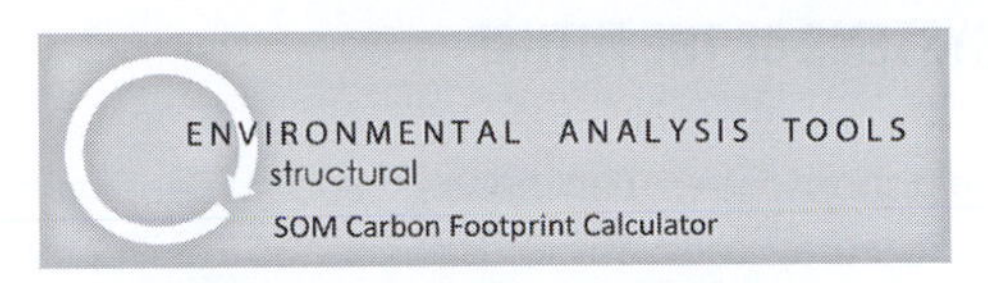

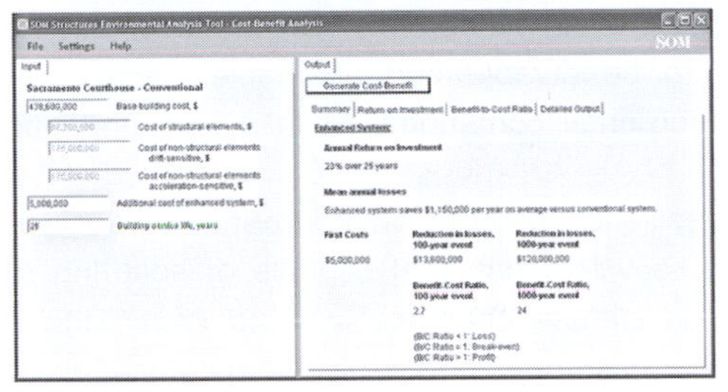

Enhanced system saves $1,150,000 per year on average versus conventional system

Cost-Benefit Analysis

Sacramento Criminal Courthouse

Parameters:

- Base Building Cost: $ 438.6 million
- Enhanced System Estimated First Cost: $ 5.0 (1.1%)

Return on Investment:

- Expected Annual Loss Benefit = $1,150,000 / yr
  - ROI = 23% over 25yrs

Benefit/Cost Ratio, 100-yr event:

- Reduction in loss = $13.6 million
  - B/C Ratio = 2.7

Benefit/Cost Ratio, 1000-yr event:

- Reduction in loss = $120.0 million
  - B/C Ratio = 24

Cost–Benefit Analysis Considering Seismic Isolation (US Patent 8256173)

The Environmental Analysis Tool™ considers the first cost of these systems and performs an analysis of anticipated damage and costs over the structure's specified life to calculate the cost–benefit ratios. The cost–benefit analysis considers the annual rate of return, mean annual loss savings, and first costs.

In addition, cost–benefit ratios are calculated for 100 and 1000 year seismic events. A cost–benefit ratio greater than 1 indicates a profitable investment. For instance, the State of California considers a minimum cost–benefit ratio of 1.5 to be required for the potential of a 100 year seismic event over a service life of 25 years.

Most insurance companies require a PML analysis. This analysis represents the total expected loss due to damage as a percentage of total cost of the building (including all components). The higher the PML, the greater the damage and the greater the expected cost to repair the damage. For code-compliant buildings the PML can be expected to range from 10 to 20; older structures designed to previous codes may have a PML of 20 or above; buildings with enhanced seismic systems such as seismic isolation, will likely have a PML of 10 or less with values as low as 2 to 4.

12.4 SUSTAINABLE URBAN SYSTEMS OF THE FUTURE

The development of new neighborhoods, campuses, and cities, and maintenance of existing infrastructure, calls for a balance of resiliency, self-sufficiency, and regeneration. To avoid the depletion of natural resources, structures must be designed to be durable and adaptable: capable of coexisting with imposed environmental conditions while accommodating changes in use.

Beyond sustainability, resiliency leads to environmentally sensitive buildings perhaps consisting of recycled materials capable of adapting to conditions of the future such as climate change. Systems within these buildings require a design ethos based on performance in which every component has multiple uses: structural systems capable of heating and cooling, exterior wall systems capable of absorbing and storing energy, and building systems capable of operating with site-based water collection, power generation, and distribution. Buildings should be completely self-sufficient, not relying on their neighbors. Advances in energy storage will be provided to bridge periods of limited or lack of solar power, while on-site water reclamation, purification, and reuse will reduce demand on our most important resource.

Morphogenetic planning of the future will consider weighted parameters for design beyond individual buildings. Form, building material, embedded and operational carbon, daylight, use efficiency, site placement, and other important parameters will be considered even on the district or city scale at early conceptual stages. The abundance of data will inform sensory fields, where magnitude and direction of oncoming environmental changes can be anticipated, gathered, reported back, and used to inform optimal performance of structures. Structures will become self-reflective, capable of undergoing state changes of materials where component properties can be temporary altered to efficiently resist abnormalities in loading. Rheological systems will use the flow of materials triggered by advanced analysis models interconnected with the building and the region's sensory field. Structural systems and all building components will be designed to behave naturally in the environment, free of the potential for damage due to extreme conditions such as seismicity.

Ultimately structures will exist in a true state of equilibrium in which umbilical reliance on services from other sources is eliminated and regeneration of resources is possible; structures will contribute to the environment rather than challenge it. This goal will only be achieved through innovative processes of collaboration, invention, and integration.

12.4.1 Sensory Fields

The response of structures within the urban context is most significantly affected by imposed load, material characteristics including stiffness, mass, shape, and the connection to the earth. Mathematical models accurately depict the characteristics of these structures; however, loading magnitudes and directionalities are typically enveloped in design leading to conservative anticipated structural demands and consequently a waste of materials. Analytical models are typically standalone and are systematically subjected to different imposed enveloped force conditions. Defining the precise magnitude and direction of imposed loads results in optimal performance and minimal materials are required for this response.

Biological-based sensory systems used to monitor specific site conditions on a district or city scale can inform buildings systems of a required response to imposed demand from natural events. A deterministic assessment of seismic ground motions, for instance, results in a definition of force vectors and energy associated with an imminent earthquake. A similar technique could be used to evaluate wind conditions. A field of sensors including accelerometers and anemometers could pinpoint natural force flows on a district or city-wide grid, and these sensors could be linked to advanced analytical building models.

Interactively linking mathematical models to information sources fed from a sensory field would enable structures to respond intelligently, perhaps systematically changing internal properties or triggering active mechanisms. Material or motion sensor systems would be placed within structures to evaluate real-time behavior under load, correlating actual and predicted performance, and also providing a stress state map of materials that could have experienced plastic stress states or permanent deformations.

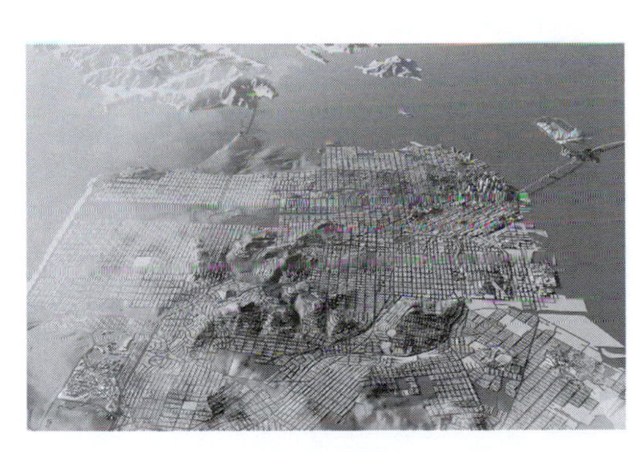

Digital Image of San Francisco with the Concept of a Mapped Sensory Field

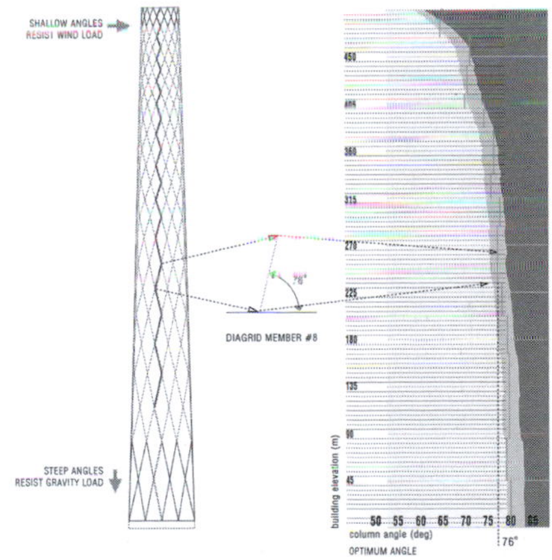

Advanced Analytical Building Model

12.4.2 Self-Reflection

Lengthening a structure's characteristic (fundamental) period results in less demand from ground motions caused by earthquakes while maintaining material elasticity, and is essential to achieving the highest performance with minimum damage and greatest chance of the building's ability to return to service following an extreme event. Seismic isolation is a technique of artificially lengthening a structure's period by decoupling the building's base from the ground. The rotation or "fusing" of joints accomplishes the same goal. For example, cast concrete roof truss elements incorporating steel pinned connections at the R&D buildings of the five-story tall Huawei Shanghai Technology Headquarters unify the multiple buildings into a complex with a consistent and lasting identity. The trusses are designed to separate or "fuse" at their apexes in major earthquakes breaking the modules linked across the atria into smaller, simpler, stable structures with cantilevered trusses—thereby reducing potential damage.

The fused pin detail used for the Huawei Project can be applied to structural systems that must be protected in tall buildings. For instance, steel outrigger systems could incorporate this system to allow for fusing to occur at extreme loads, protecting primary truss members from permanent damage.

Similar to the way individual plants within the same species react to different placement in the environment with variations in growth patterns or luminary control devices that determine required light levels for supplemental lighting, structures should be designed to be environmentally reactive, dynamic, and self-reflective. Sensory information flows would inform structures of anticipated demand and allow for an interactive response.

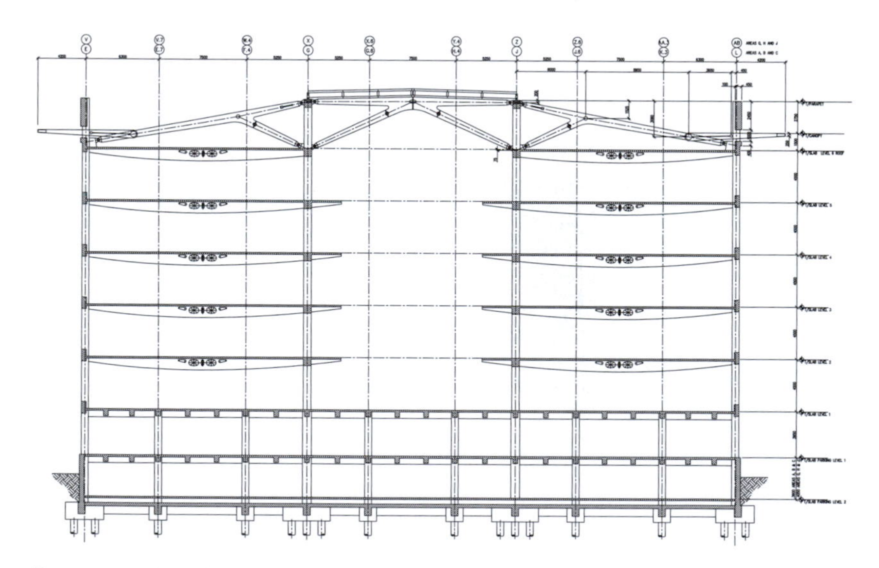

Transverse section of Shanghai Huawei Technologies Corporate Campus Building, Shanghai, China

Shanghai Huawei Technologies Corporate Campus, Shanghai, China

Fused Cast-in-Place Concrete Roof Trusses, Shanghai Huawei Technologies Corporate Campus, Shanghai, China

Analytical mathematic models for the structure would be directly linked to imposed magnitude and direction of load, informing structures of appropriate next steps in response. These next steps could include activating strategically placed behavior controlling devices or mechanisms that alter joint or base connections.

For instance, pneumatic dampers that incorporate compressed air or visco-elastic fluid would be activated and tuned interactively by interactively correlating the actual dynamic response with the predicted mathematical dynamic response. Damper activation would be introduced sequentially focused in critical areas of the structure where its participation could be most beneficial. Important research was conducted at the Technical University of Berlin where high-strength carbon ribbon was used for a long span pedestrian bridge. Because the structure was highly sensitive to walking induced vibration, dampers were introduced into the handrails. These compressed-air, pneumatic dampers were able to immediately dampen the motion when activated.

Damper Activation Research, Technical University Berlin, Germany

In the case of strong ground motions, a more effective and sophisticated response would be to create complete separation from the input source and the structure. Temporary levitation created by electromagnetic flow or temporary air cushions similar to the technology used by Poma Otis for the Skymetro train line in Zurich would provide frictionless seismic isolation by creating a separation of the superstructure from its foundations. Soil liquefaction under superstructures is typically mitigated by deep foundation systems. Instead, perhaps this behavior should be encouraged by advantageously utilizing the subgrade rheology by using the ground motions to reduce soil shear strengths and, therefore, the capability of force transmission between the ground and structure. A perimeter membrane restraining system could be used to limit the possibility of uncontrollable property changes. The viscosity change in the soil would be similar to that of ketchup where its liquid property changes through shear thinning achieved through mechanical agitation.

The fixity of certain joints within the structure could also be modified on-demand. For instance, if clamping forces could be temporarily relieved, the stiffness of the frames would be reduced, and the natural periods of vibration lengthened with a reduction in attracted inertial loads. The reduction of clamping forces could be achieved by introducing energy in the form of heat into these joints through bolt fastenings where lengths would be temporarily increased through heat expansion/bolt elongation.

To control joint behavior and allow for structural re-centering, joint sinews could be introduced using counteractive high strength strands or shape memory alloys such as nickel-titanium (NiTi) where the inherent elastic properties of the materials are used or heat is applied, changing the material from an austenite to martensite state after cooling.

Since the fundamental period of vibration of structures is directly proportional to the square root of the reciprocal of mass, reducing a structure's mass while maintaining the same lateral stiffness increases the period and its willingness to attract force from ground motions. It is known that approximately 25% of the concrete placed in conventionally constructed buildings is not only unnecessary for strength, but increases mass and demand on vertical load-carrying elements such as walls and columns. For example, most of the concrete placed in center spans of structures is not required and is placed because of the ease of construction. In addition, the environment is becoming overwrought with waste materials that do not decompose and are not recyclable. Materials such as lightweight, waste plastics and polystyrene could be strategically substituted for redundant concrete, benefiting the environment, as well as reducing mass. The Sustainable Form Inclusion System™ (SFIS), originally conceived to create air voids in structures by placing capped, empty plastic beverage containers into structural systems, achieves these goals.

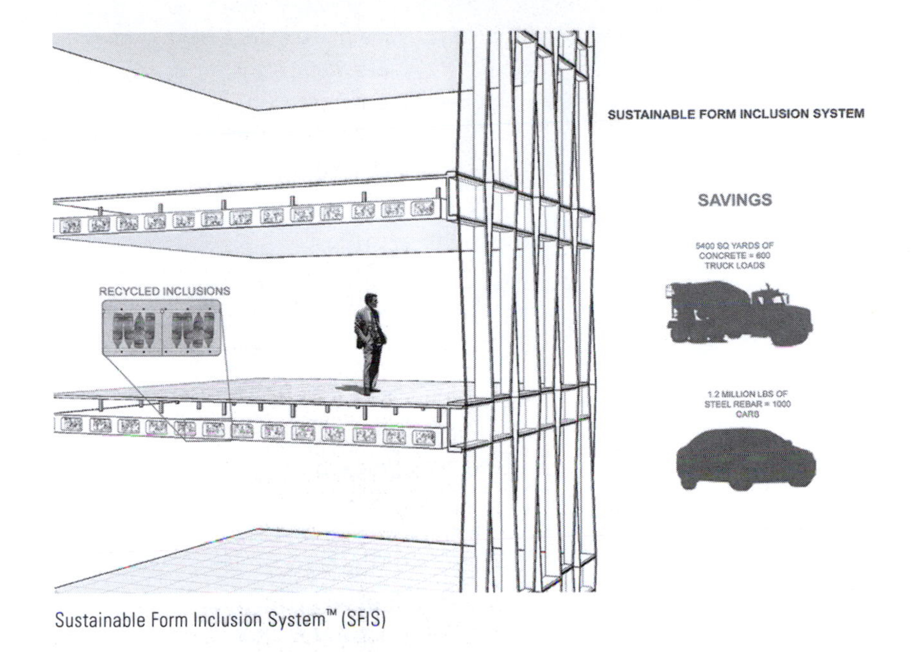

Sustainable Form Inclusion System™ (SFIS)

More practically, the system can utilize "bricks" comprising ground and formed plastics or waste Styrofoam cast into a lightweight mortar. Environmental responsibility could be further achieved by using "zero-cement concrete" products such as GREENCEM where cement is eliminated and waste blast furnace slag used as a replacement in combination with special binding agents.

12.4.3 Morphogenetic Planning for the Future

Evaluating multi-variable parametric building models on a district or city scale can be used to help inform best planning strategies for the future. The Parametric City Model (PCM) combines the weighted importance of form, structure, embedded carbon, and efficiency of space use while considering orientation, including exposure to daylight, and solar gain. This model is interfaced with programs such as Grasshopper, used to define geometry, combined with tools such as Galapagos and Karumba where genetic algorithms and structure can be defined. The model accesses a database of hundreds of previously designed and built structures. The recorded data includes structural requirements relative to height, material type, height, and site location (seismicity and wind conditions) along with space requirements for building systems such as vertical transportation and mechanical systems. The PCM can evaluate the embodied carbon impact of construction considering material types (steel, concrete, wood, masonry, etc.), fabrication and transportation of these materials, construction time and required equipment, the number of

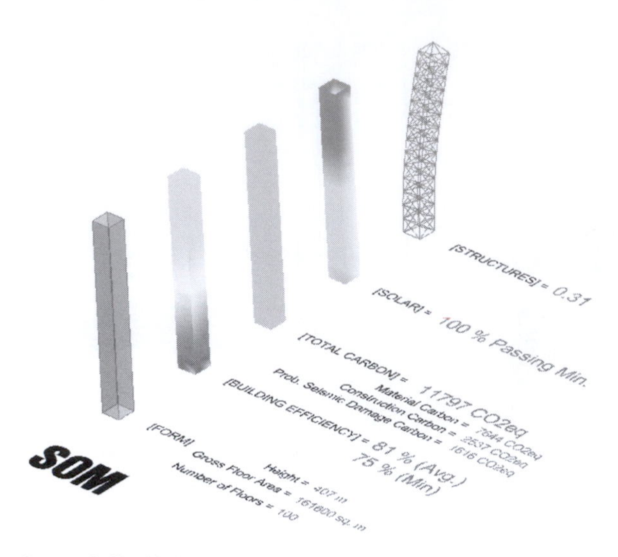

Parametric City Model for the Evaluation of Multiple Design Variables

construction workers and their transportation to and from the site. With the requirements for structure and mechanical systems known, net available space can be evaluated for commercial value based on location within the structure (i.e. floor level), access to daylight, and views. The model is also capable of evaluating the environmental and financial benefit of incorporating advanced seismic systems into structures through the reduction of life cycle carbon, anticipated damage over time, and the cost–benefit of addressing those risks at the time of construction. For slender structures or structures with complex geometries, parameters can be interactively evaluated considering the advantages of interlinkages or other geometry modifications.

These models can be translated into more sophisticated structural analyses to determine the placement of structure where least energy is expended when work is done to resist load. In minimizing the energy, forces and deformations should be distributed as evenly as possible throughout the structure through a synergetic placement of material. Forces will flow through the easiest and shortest load path natural to the structural form. Topographical optimization techniques are used to map the structural response to define the most efficient placement of materials.

Parametric City Modeling can be done on a district or city scale. The model can be calibrated to indicate areas where the net usable area falls below a target minimum, in this case 75% of the gross floor area (darker areas near the bottoms of the towers). Seventy five percent of the gross area is the typical minimum target for buildings that could become financially viable. This net area is the resultant area after structure, elevators, stairs, and services including mechanical/electrical/plumbing systems have been accounted for. This Parametric Modeling System is particularly important for the building owner and they develop proforma for the rental or sale of developed properties.

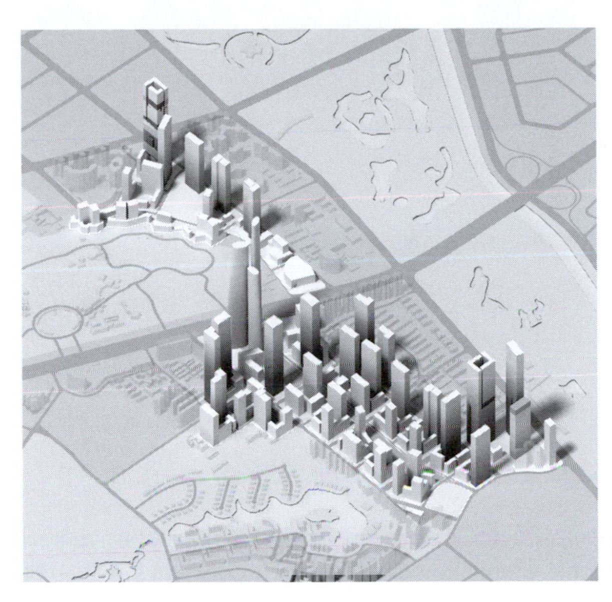

Parametric City Model for the Evaluation of Net Usable Area in Buildings

On a district or city scale, environmental impacts of planning can be interactively evaluated based on proposed or anticipated building material type, geometry, and site conditions. The type of use plays a significant role in the overall plan since the requirements for building systems and structure vary when comparing, for example, office, residential, and mixed use occupancies.

12.4.4 Rheological Buildings for the Future

The envelope enclosure for structures represents the single greatest opportunity to consider flow and interaction between architectural, structural, and building service systems. Hundreds of millions of square feet of occupied area are enclosed each year with systems that essentially provide protection from the elements, safe occupancy, and internal comfort. A closed loop structural system integrated with the exterior wall and roof system that includes liquid-filled structural elements could provide a thermal radiator that, when solar heated during the day, could be used for building service systems such as hot water supply or heat for occupied spaces during the evening hours. A solar collection system could be integrated into the network and incorporated into double wall systems where it can be used to heat the internal cavity in cold climates.

Transparent photovoltaic cells could be introduced into the glass and spandrel areas to further capture the energy of the sun. When storing fluid in structural systems of great height, pressures within the networked vessel become very large. With this level of pressure, water supply systems to the

Rendering of Poly International Plaza, Beijing, China

structure or to neighboring structures of lesser height could easily be supplied without requiring additional energy to move the water. Constant low flow or a material with a low freezing point moving through these systems would prevent the liquid from freezing.

The Poly International Tower includes a structure type that could include liquids in the future. Liquid within the networked system would control motion with fluid flow acting to dampen the structure when subjected to lateral loads due to wind and earthquake events. If compartmentalized at key moments during service through capped cells, liquids at high pressure could add significantly to the axial stiffness and stability of structural members subjected to compression (pressurized water hose effect). Combining ultra-high strength tensile materials such as carbon fiber fabricated into

Poly International Plaza Under Construction, Beijing, China

State Change—Solid to Liquid or Liquid to Solid

closed circular forms where loads are primarily resisted by hoop stress with the liquids under ultra-high compressive stress would likely result in greatest efficiency in resisting applied load.

The concept of flow can be further developed into structures that are interactively monitored for movement. Through the measurement of imposed accelerations due to ground motions or wind, structures could respond by changing the state of the liquid within the system. For instance, the structure could use endothermic reactions to change liquids to solids within the closed network. Sensor devices could inform structural elements of imminent demand and initiate a state change in liquids that would be subjected to high compressive loads where buckling could occur. For example, water within the system could be frozen for additional structural rigidity.

Magnetorheological or electrorheological (ER) fluids could be used to change the viscosity and, therefore, stiffness of closed vessels in addition to changing damping characteristics. When subjected to a magnetic field, magnetorheological fluids greatly increase their apparent viscosity and can become viscoelastic solids. When subjected to an electrical field, ER fluids can reversibly change their apparent viscosity, quickly transitioning from a liquid to a gel and back again.

12.5 THE SELF-SUFFICIENT TOWER

Perhaps in the not so distant future, towers or at least a district of towers will be designed and constructed to not only achieve net-zero energy as defined by the American Institute of Architects (AIA) 2030 Challenge, but also become self-sufficient. This goal would remove all conventionally implemented sustaining utility service connection such as power, fuel for heating and cooling, water, telecommunications, and waste. Ideally, the tower would perform beyond self-sufficiency and would contribute to the district or city by distributing excess power or supplying clean water.

Ultimately structures will exist in a true state of equilibrium where umbilical reliance on services from other sources is eliminated and regeneration of resources is possible; buildings would become self-sufficient or perhaps would contribute to the environment rather than challenge it.

12.5.1 Structural Systems with Multiple Purposes

Structures in tall buildings of the future should be designed to have two or more purposes. For instance, the structural system could be a conduit for fluids that can be used for heating and cooling buildings. Structures can be carefully integrated into the exterior wall systems where superficial enclosure elements are eliminated.

Devices could be used to open and close controlling direct sunlight. This radiator concept could be used to heat the fluid in the winter and cool the fluid in the summer. If water is used in these systems, then it also can be used for anything from fire protection to irrigation to controlling building movements as a damping system. This is not an original concept and has been successfully used in the past. For instance, the Gartner brothers incorporated this system into the design studio space at their exterior wall fabrication plant in Gundelfingen, Germany, over 50 years ago. New technologies such as chilled beams show new promise for such a system.

The 150 m (292 ft) tall Poly International Plaza, Beijing, provides a basis for consideration of such a system. Although the open pipe system was ultimately filled with concrete to increase seismic resiliency, this fully welded system could have been a conduit for a liquid such as water.

Steel Pipe Construction and Exterior Wall System, Poly International Plaza, Beijing, China

Steel Pipe Construction and Exterior Wall System, Poly International Plaza, Beijing, China

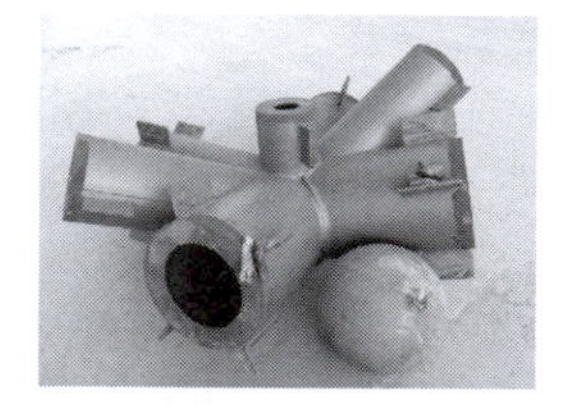

Prefabricated Pipe Joint, Poly International Plaza, Beijing, China

12.5.2 Advancements in Self-Sufficiency

12.5.2.1 Designing for Net-Zero Energy

The 71 story, 309 m (1013 ft) tall, 213,700 square meter (2,300,000 square foot) Pearl River Tower, Guangzhou, China is one of the most energy efficient super-tall buildings ever constructed. Designed to perform as a net-zero energy building, the structure integrates solar panels, a double skin curtain wall, a chilled beam mechanical system, an underfloor air distribution system, and daylight harvesting used to project light deep into occupied spaces. With the combined benefit of these integrated systems, 58% percent of the energy use was reduced related to cooling, pumps, fans, and lightning compared to a similar building meeting current building code requirements.

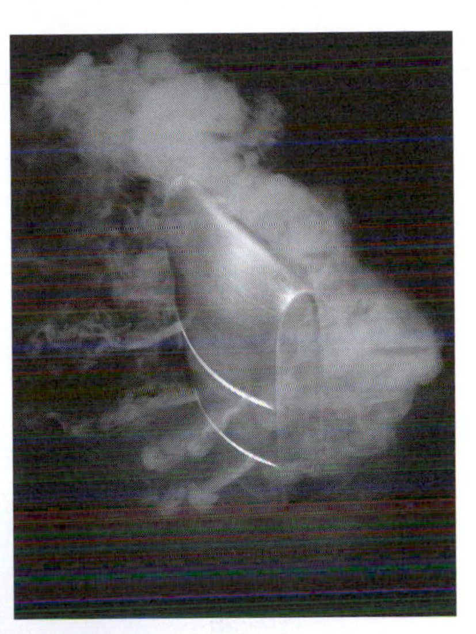

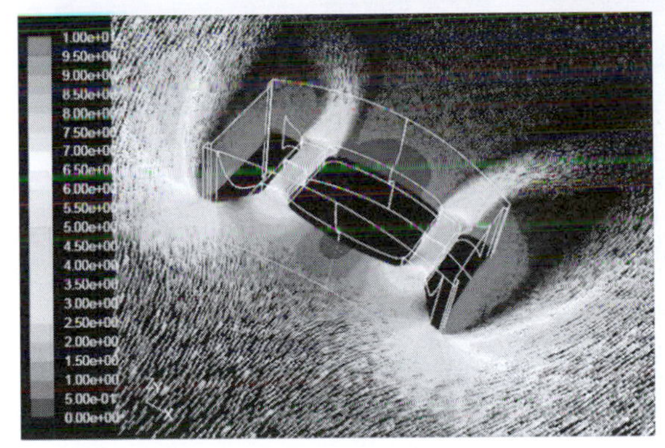

CLOCKWISE FROM TOP LEFT

Pearl River Tower, Guangzhou, China

Wind Flow Test, Pearl River Tower, Guangzhou, China

Computational Fluid Dynamics/ Wind Flow Analysis, Pearl River Tower, Guangzhou, China

Exterior Wall

Pearl River Tower's design incorporates a dynamic high performance building envelope that provides superior thermal performance, controls solar loads, and optimizes the transmittance of daylight into interior spaces. The east and west elevations use a unitized frit glass outfitted with external shades. The external glass is double layer, low-E high performance glass, which provides the best balance between low heat gain and high transparency.

A photovoltaic system is strategically integrated into the external shading system located on the east and west elevations in order to capture energy from the sun and protect the tower from solar gain.

The performance is heavily impacted by the north and south elevations. In section, these façades break down into a double wall, 300 mm unitized system with a 240 mm ventilated cavity that separates two layers of glazing. A low-E coated, insulating glass unit forms the exterior layer, and a single monolithic glazed panel forms the interior layer.

Mechanical

The radiant ceiling cooling system delivers sensible cooling directly to the space, decoupling the sensible cooling load from the latent cooling load. The radiant cooling panel system is combined with direct outdoor air systems (DOAS) and underfloor air delivery that provides improved air change effectiveness and indoor air quality.

Photovoltaic Panels Combined with
Exterior Wall Shading System,
Pearl River Tower, Guangzhou, China

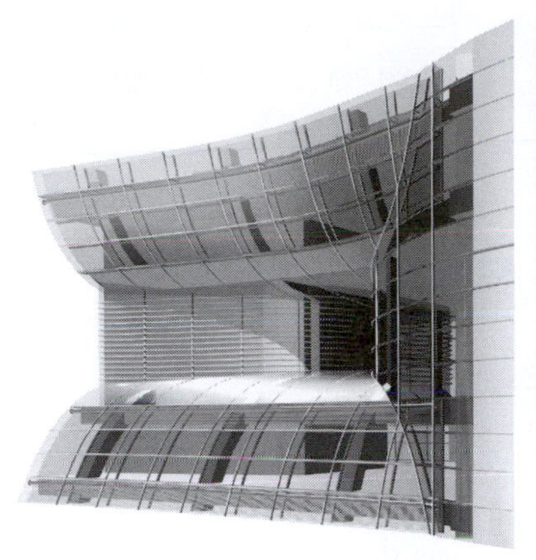

Shaped Wind Slot, Pearl River Tower, Guangzhou, China

Wind Turbine Located in Shaped Wind Slot, Pearl River Tower, Guangzhou, China

Daylight

Pearl River Tower uses daylighting responsive controls in conjunction with the ventilated cavity exterior wall system to automatically reduce artificial light levels in response to increased available daylight.

Power

The building's signature form is sculpted to allow winds to pass through the structure while increasing wind velocities through a gradual narrowing of orifices. These increased wind flow velocities are used to drive large vertical axis wind turbines and create power for the building. As a result of the precisely shaped openings, speeds are magnified up to 2.5 times.

12.5.2.2 Performance Beyond Net-Zero Energy

Understanding and utilizing the tall tower's environmental conditions is key to self-sufficiency. Winds at the site can be used to not only generate power, but also control behavior or both. Holes introduced into the tower along the height of the building can also:

1. Allow winds to path through the building lessening the surface area subjected to winds while minimizing across-wind dynamic effects.
2. Allow power to be generated at each opening location. A reduction in the opening diameter within the structure increases wind velocity and power generated.
3. Incorporate an airfoil concept where forces can be developed to counteract overturning by generating upward forces on the leeward side of the structure.
4. Incorporate windcatchers to funnel air into cooling systems where air is moved over a reservoir of water.

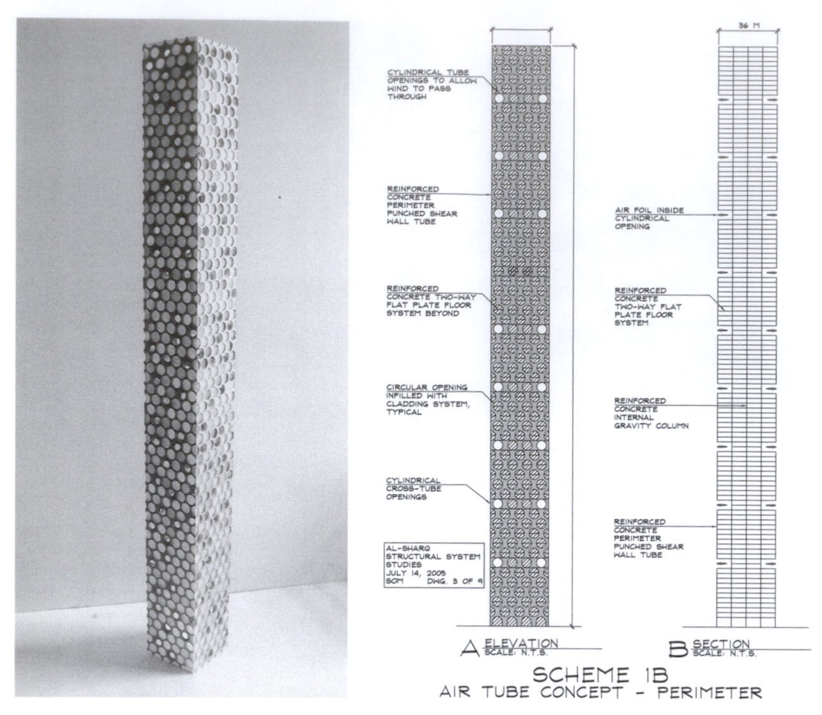

High-Rise Tower with Air Tube Concept

Elevation—Air Tube Concepts—Openings and Air Foils

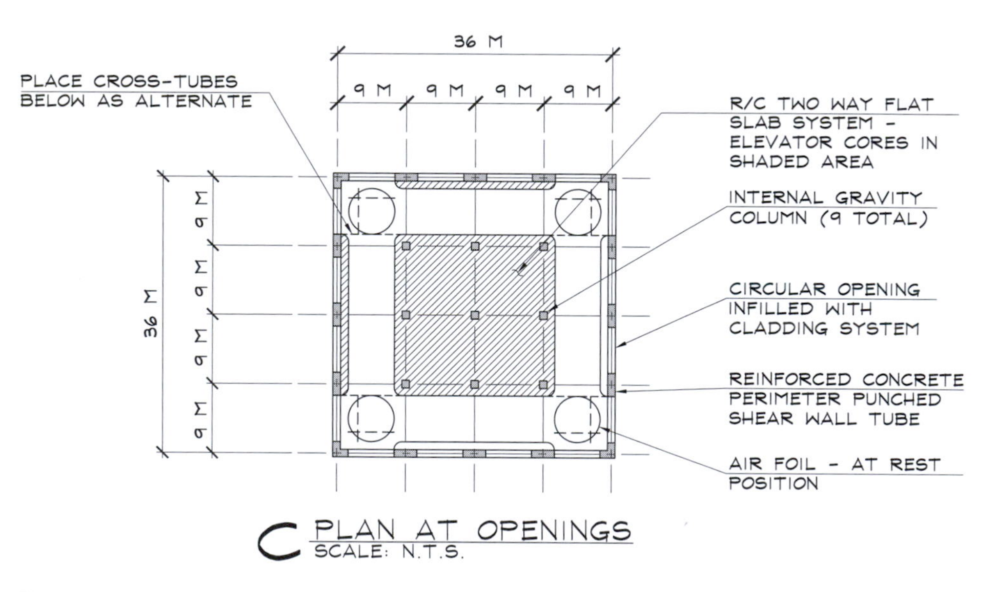

Plan—Air Tube Concept—Openings and Air Foils

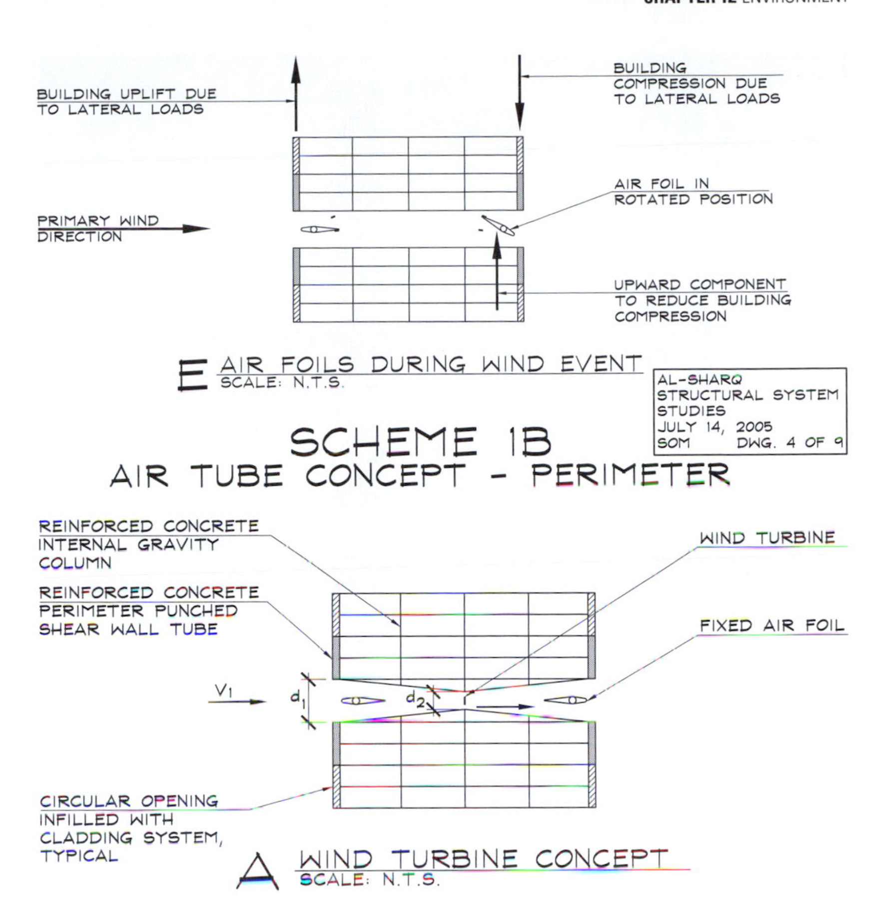

(top) Partial Elevation—Air Foils Activated in Wind Event

(bottom) Partial Elevation—Air Foil and Wind Turbine Concept

12.5.3 Optimized Systems that Lead to Self-Sufficiency and Resiliency

Optimization of systems can be performed for most components within the tall building. For the mechanical systems this optimization may be related to the efficiency of systems. For the exterior wall systems, it may be related to absorbing energy while controlling heat gain. For the structure, material selectively distributed throughout the building resulting in an efficient use of material will minimize embodied carbon. The design of the structure is usually optimized to achieve the maximum stiffness of the building, which is typically measured by the compliance of the lateral system, or by the tip displacement of the building itself. Therefore, the resulting design is the stiffest structure that can be obtained for the volume fraction of material considered and satisfies the performance requirements under gravity and lateral loads (wind and seismic).

Optimized Braced Frame, CITIC Financial Center, Shenzhen, China

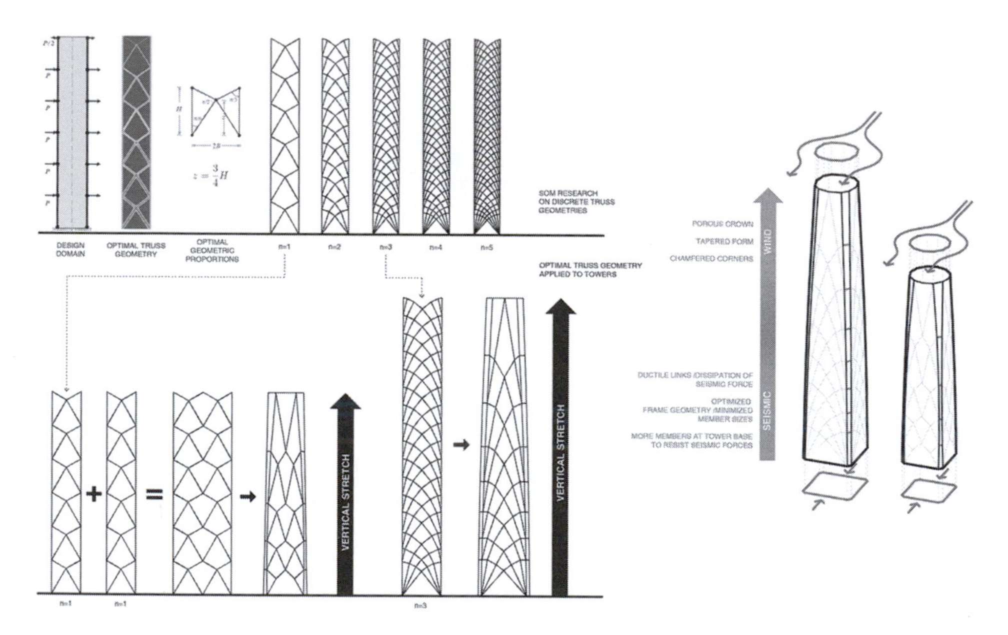

Evolution of Theory to Concept for Optimized Braced Frame Structural System, CITIC Financial Center, Shenzhen, China

12.5.3.1 Evolution of Theory to Concept for Structural System

The optimal geometry is usually derived under the assumption of elastic material behavior, which is accurate for wind loads and earthquakes with a relatively low return period. However, when earthquakes with a longer return period are considered, the structure will exceed the elastic limit and ductile elements should be incorporated in the stiff lateral system to ensure energy dissipation with consequent reduction of the inertial loads on the structure.

The seismic fuses could consider use of a conventional ductile steel link similar to those used with eccentric braced frames, or could use a modified link–fuse joint that is designed to protect the base building structure from permanent damage. The conventional ductile steel link is designed to yield during a significant seismic event where the link-fuse uses a clamped pin connection, dissipating energy through sliding of the joint.

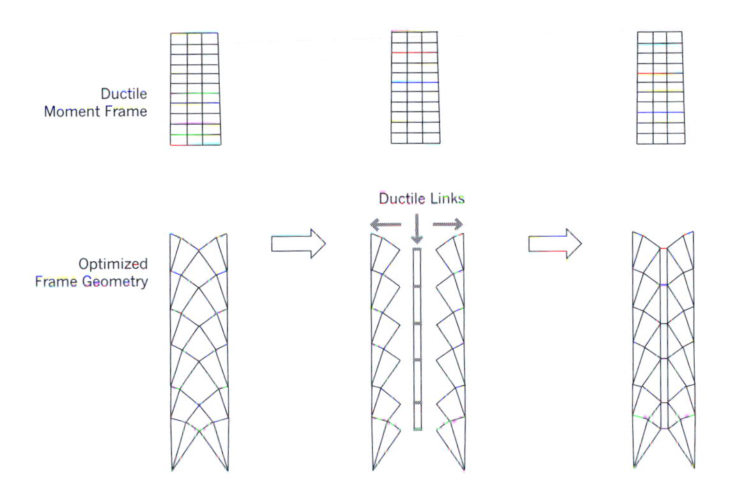

Optimized Braced Frames Interconnected with Seismic Fuses,
CITIC Financial Center, Shenzhen, China

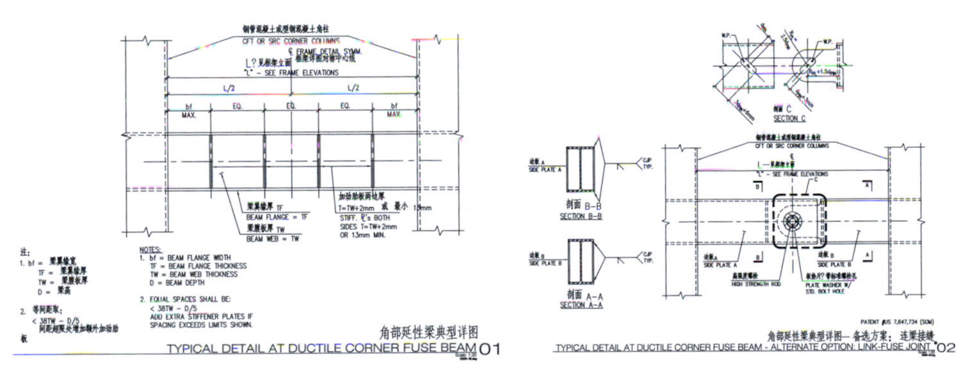

Conventional Ductile Steel Seismic Link

Futuristic Link-Fuse Seismic Joint for Ductile Links

Rendering of Citic Financial Center Towers—
Office, Hotel, and Residential Uses

Conceptually, the systems mimic plant behavior where trunks carry water vertically to branches and leaves, with active movements from the ground to the top. For this structure the principles are the same, only forces replace water and active flow from the ground is replaced with resistive forces originating in the superstructure, the seismic mass of floor framing systems, or winds applied to each floor diagram passing into the braced frame down to the foundations. Shear and bending moments increase as these forces near the ground and so does the density of the structure. The structure could have multiple purposes and would be resilient, capable of acting elastically with minimum damage in strong earthquakes with optimal performance for wind and frequent seismic events.

GLOSSARY

Acceleration—the rate of change of velocity over time. In tall building design, wind-induced acceleration is usually expressed in milli-gs or thousandths of the acceleration of gravity.

Acceptance criteria—criteria that establish demand to capacity limits, story drift limits, and stress limits on the structure.

Across-wind motion—motion of the structure in the direction of the applied wind load. Also known as lift.

Aerodynamics—air flow around a solid object of various shapes.

Aero-elastic structural modeling—model used in wind tunnel testing that incorporates actual building structure properties including shape, mass, stiffness, and damping. Used to directly measure displacements and accelerations during the wind tunnel testing.

Allowable stress design—the capacity of a structural material is based on the allowable load that the material could resist without permanent or plastic deformation. Stresses due to service loads do not exceed the elastic limit of the material. Also known as permissible stress design.

Along-wind motion—motion of the structure perpendicular to the applied wind load. Also known as drag.

Aspect ratio—the ratio of the height of a structure relative to its smallest plan dimension at the base.

Axial load—Load applied along the axis of a member or system.

Axial stiffness—stiffness of axial load resisting elements within the structure—important to the effectiveness of outrigger trusses interconnected to the core and columns.

Bamboo—a plant having a mathematically predictable growth pattern, consistent structural properties, and fast natural growth.

Base ten logarithm—the logarithm of a given number to a given base (10) is the power or exponent to which the base must be raised to produce the number.

Beam-to-column—a joint that describes the connection of a beam to a column.

Belleville washers—compressible washers used to maintain bolt tension after sliding in a joint has occurred.

Belt shear wall-stayed mast—a central reinforced concrete shear wall is interconnected with perimeter columns or frames with outrigger trusses or walls.

Belt truss—a steel truss used to tie perimeter columns together and typically used to transfer load from outrigger trusses to perimeter columns.

Bending moment distribution—applied bending moments along the height of the structure.

Binary digits—the digits of 0 and 1.

Boundary conditions—support conditions of the structure that must be coincidently considered with the application of load.

Building service systems—systems that typically include mechanical, electrical, plumbing, and telecommunications components required for a tall building.

Bundled frame tube—a bundling or collection of tubular frames that typically pass through the interior of the structure—belt trusses are typically used at building geometry transitions.

Bundled steel tubular frame—an assembly and interconnection of individual tubular structural steel frames.
Buttressed core—a strong central core used to anchor three wings to form a tripod shape in plan. Each wing is buttressed to the other two wings.
Cellular concept—placing frames in plan to form the geometry of cells in the interior of the structure.
Chevron or k-braced truss—a steel truss with a geometry that is in the shape of a K with a horizontal orientation—typically used in core or outrigger truss applications.
Clad structural skeletons—systems such as frames or walls that provide primary support for the building structure that are clad or covered with exterior curtain wall systems.
Code-defined—requirements specifically defined by governing code for design.
Column tributary area—floor-framing area supported by vertical columns.
Composite—in tall building construction usually refers to the combination of steel and reinforced concrete.
Composite frame—a frame that includes columns and beam comprised of both structural steel and reinforced concrete.
Compressible (soil) layers—soil layer susceptible to displacement when subjected to compression loads.
Concentric braced frames—diagonal frames in a structure that share a common joint without any offset.
Concrete diagonal braced frame—a braced frame in reinforced concrete where panels are infilled between frame columns to form diagonal patterns.
Cost–benefit analysis of structures—the evaluation of considering first cost relative to long-term costs of a structural system of device designed to provide better performance.
Creep—the tendency of material to deform permanently under the influence of stress.
Creep effects—the tendency of a solid material (soil in this case) to slowly move or deform permanently under the influence of stresses due to sustained loads.
Culm—the stalk or trunk of bamboo—typically hollow except at nodal or diaphragm locations.
Curtain wall system—an exterior wall system used to enclose the structure and protect against natural elements such as wind, moisture, solar, etc. Wall system does not provide structural support for the base building frame.
Cyclic behavioral—structures subjected to loads or displacements that start in one direction then reverse to the other.
Damping ratio—an effect that acts to reduce the amplitude of oscillation. Usually referred to as the percent of critical damping. Building elements that provide damping include non-structural components such as partitions, ceilings etc., the structure itself, aerodynamics, and so on.
Dead load—load that is constant over time typically referring to the structure's self-weight.
Design criteria—criteria that describe structural system, anticipated ground motions, code references, and performance objectives.
Design earthquake (DE)—having an intensity 2/3 x MCE typically having a probability of exceedance of 10% in 50 years or a return period of 475 years.
Design loads—loads used for the design of a tall building that may be sustained or transient. These loads include gravity, wind, seismic, temperature, snow, among others accounting for specific building components and systems.

Design spectral response acceleration—equal to 2/3 of maximum considered spectral response acceleration for a site resulting from a 475-year seismic event.

Detailing—special structural design and construction requirements used to manage applied loads allowing primary structural members to fully utilize their strength and stiffness properties. Typically associated with connections within a building frame and are critical in the design of seismically resistant structures.

Diagonal braces—braces within the structure placed diagonally to resist loads.

Diagonal mesh tube frame—a reinforced concrete frame made up of smaller, more repetitive diagonal members—typically located at the perimeter of the structure and has cellular characteristics.

Direct positive pressure (windward faces)—pressure created from the velocity of wind on a surfaces or surfaces perpendicular to the direction of the wind.

Direct tension indicators—washers used to predict bolt tension through the deformation of tabs incorporated into the washers.

Ductility—the ability of the structural system to dissipate energy without failure (ie. structural steel permanently deforming through bending with fracture).

Dynamic—having motion, applied forces result in motion. With structures exposed to wind forces it is the response of the structure caused by changes in fluid (air) flow.

Dynamic properties—natural properties of the structure that include mass and stiffness.

Earthquake force—force experienced by structure due to ground motions from seismic events.

Eccentrically braced frame—a braced frame where work points of diagonals are not common at horizontal members.

Effective seismic weight—dead and live load specifically considered in seismic analysis.

Elastic drift—displacement of the structure when applied loads do not cause permanent deformations and allow the structure to return to its original undeformed position.

Elastic shortening—the reduction of last in a structural element due to load. The element returns to its undeformed shape after the load is removed.

Emergence theory—emergence or self-organization is the interaction between simple common elements having singular and common characteristics, each functioning according to its own simple rules, resulting in complex behavior, without an obvious central controlling force.

Enhanced performance objectives—objectives that lead to better structural performance in an earthquake including considerations beyond code for ground motion input, components, or systems.

Enhanced seismic systems—structural systems that introduce technologies that lead to better performing structures in seismic events where higher performance and less damage is achieved.

Environmental Analysis Tool™—a calculator used to determine the carbon footprint, cost–benefit, and life-cycle of a structure.

Exterior diagonal tube—diagonal structural members incorporate into a tubular system at the perimeter of the structure.

Exterior frame tube/tubular frame/framed tube—a frame with closely spaced columns having a plan spacing typically similar to the floor-to-floor height.

Faying surface—an interface surface between two or more materials that has a characteristic coefficient of friction.

Fibonacci Sequence—rooted in logic formations created by binary numbers; it is a series of numbers starting with 0 and 1 where the subsequent number in the sequence is equal to the sum of the previous two numbers.
Five Ages of the Skyscraper—five ages or periods of development of the skyscraper starting after the Great Chicago Fire with the First Chicago School of architecture, then through the use of new materials like structural steel, modernism, postmodernism, and finally the new age where performance of structures, types of materials, construction practices, embodied energy, and a holistic approach to the integrated building are considered.
Floor-to-floor height—the overall vertical dimension of the structure from floor slab at one level relative to the floor slab and a neighboring level.
Force distribution in indeterminate structures—the distribution of forces due to gravity or lateral loads (ie. wind or seismic) within a structural frame. Indeterminate structures are those that cannot be defined by the fundamentals of engineering statics. Statics refers to a state of equilibrium where no bodies are in motion and forces are offset or counterbalanced. The stiffness of members must be considered in defining the distribution of forces within the structure.
Force flow—the internal flow of force throughout a structure.
Force–balance structural modeling—block model generally replicating building shapes placed in expected wind environment and boundary layer that includes neighboring terrain/building structures. Bending moments due to applied wind loads are measured and applied to an analytically model of the structure that includes structural member stiffness, mass, and expected damping.
Form and response—geometry considerations for the tall building that leads to the most favorable response. Tapering of form and placement of structure is key to reducing wind and seismic demand.
Fragility curves—mathematical curves defining performance of structural system when subjected to large displacements and loads—used for risk assessments of structures.
Frame–shear wall—a combination of a frame and shear wall.
Frequent or service level earthquake—an earthquake ground shaking having a 50% probability of exceedance in 30 years or a return period of 43 years.
Friction-type—a connection that relies on friction to determine capacity.
Fundamental period of vibration—the longest period (or lowest frequency) of vibration for the structure. Also referred to as the first or primary mode of vibration.
Genetic algorithm—search technique used in computing to find exact or approximate solutions to optimization and search problems.
Gravitational acceleration—the acceleration due to gravity (32.2 ft I s2).
Gravity loads—loads due to the self-weight (based on density of the material used) of the structure or superimposed loads (dead or live) on the structure.
Gross floor area—the overall floor area within the exterior wall enclosure without consideration of typically openings (shafts, building service openings etc.) but with the subtraction of large open areas such as atrium.
Growth patterns—patterns of natural growth observed in nature having structural characteristics.
Hand calculation techniques—manual calculations without the use of computers considering the fundamental principles of physics and mathematics.
Helical formation—a geometric definition of a form based on the helix.

Higher compressive strengths—compressive strength of concrete typically achieved with increased amounts of cement in concrete mix. Determined by destructive testing of unreinforced cylinders or cubes. Typically expressed in load per area of material and typically specified/tested at 28 or 56 days after casting.

Horizontal force distribution—distribution of horizontal story forces along the height of the structure.

H-piles, precast piles, steel pipe piles—driven piles consisting of structural steel Hshaped sections, precast concrete, and open or closed end steel pipes.

Inelastic drift—displacement of the structure when applied loads cause permanent deformations.

Inertial forces—forces generated within the structure resulting from applied motion from the ground to the foundations during a seismic event.

Initial loading—loads typically associated with the dead weight of the superstructure.

International Building Code 2012 (IBC 2012)—a model building code developed by the International Code Council and has been adopted by most jurisdictions in the United States.

Inter-story drifts—relative displacement of one level of the structure relative to another due to applied load.

kPa—kilopascal.

Lateral loads—laterally applied loads typically due to temporary events such as wind or seismic activity.

Life-cycle—the estimated life of a structure given site and loading conditions.

Limit state—the application of statistics to determine the level of safety required for the design of structural members or systems.

Link-Fuse Joint™—a patented seismic frame device that uses friction fuse to dissipate energy during an earthquake—uses pins, butterfly link-beam steel plate connection, brass shims, and high strength steel bolts.

Live load—temporary loads typically of short duration. Typically includes occupancy loads or snow loads.

Load and resistance factor design (LRFD)—synonymous with limit state where the application of statistics to determine the level of safety required for the design of structural members or systems. Most commonly known for the structural steel design code that superseded allowable stress design.

Logarithmic spiral—a mathematical definition based on logarithms for geometric formations found in nature including shells, seeds, plants, spider webs, hurricanes, and galaxies.

Long-term loads—sustained loads on a structure such as dead load and permanent superimposed load such as exterior wall systems.

Mat foundations—foundation system that distributes load from the superstructure to the soil system typically over an area much larger than spread footings. Usually comprised of reinforced concrete and supports multiple vertical load carrying elements such as columns or walls.

Maximum amplitude of oscillation—is the maximum magnitude of change in the oscillating variable, ground motion.

Maximum considered earthquake (MCE)—earthquake typically has a probability of exceedance of 2% in 50 years or a return period of 24750 years.

Maximum considered spectral response acceleration—the maximum considered spectral acceleration for a site resulting from a 2475-year seismic event.

Mechanism—a structure having joints that allow moving without permanent deformation of materials.

Mega-column—column of large scale used typically used in structural systems where gravity loads are concentrated or included in an outrigger truss system where high axial load resistance is required or a combination of both.

Mega-core shear walls—large reinforced concrete cores usually located in the central core area and are designed to resist a majoring of load with or without interconnection with perimeter columns or frames through an outrigger system.

Mega-frame concept—the use of a frame that typically extends over multiple stories of the structure.

Mill scale—the surface of steel left after a typical milling process.

Mitchell Truss—the mathematical derivation by Anthony Michell defining the geometry of the perfect trusses subjected to lateral loads with two points of support.

Modal analysis—the study of the dynamic properties of structures when subjected to vibration-induced excitation.

Modular tube—a tubular frame that is included both at the perimeter of the structure and through its interior.

Modulus of elasticity—the linear relationship of stress to strain in a material. The tendency of a material to behave elastically when load is applied to it.

Moment of inertia—is a property of a cross section that can be used to predict the resistance of a structure when subjected to bending or deflection, around an axis that lies in the cross-sectional plane.

Moment resisting—a frame with fixed beam-to-column joints designed to resist lateral loads.

Morphogenetic planning—using advanced computer simulations techniques to model multiple parameters in the design of buildings, urban districts, and cities. These parameters include but are not limited to form, structure, embedded carbon, and efficiency of space use while considering orientation, including exposure to daylight and solar gain.

Natural period of vibration—the time required for a structure to complete one full cycle of vibration referring to the first, primary or dominant mode of vibration when given a horizontal (or vertical) displacement.

Net floor area (NFA)—the floor area remaining after area required for building services, elevator shaft area, structure, shafts and stairs are subtracted from the gross floor area.

Net tension—the resulting force on a member or system where tensile forces are greater than compressive forces due to axial loads or bending.

Net-zero—energy used and energy produced are equal over a time of consideration.

Non-prescriptive design—design approach that considers one or more exceptions to the building code. The approach to design must prove code equivalency is met.

Occupancy importance factor—a factor used to recognize structures with occupancies that require special considerations (for example acute care hospitals).

Optimal structural typology—conceiving of structural systems as membranes and performing optimization analyses to determine regions where greatest structural material density is required.

Optimization—analysis techniques used to calculate least material required for particular loadings and boundary conditions.

Optimized systems—structural systems designed based on energy principles where material is placed to provide optimum response load (i.e. greatest stiffness along with least material).

Oscillatory response—repetitive variation of an object in motion.
Out-of-plumb—alignment that varies from a pure vertical condition.
Outrigger truss—a steel truss used to share load between a steel shear truss or concrete shear wall core and columns to resist overturning causes by lateral loads.
Outrigger truss system—structural steel trusses used to interconnect the core of structure to the outside frame or columns to resist applied lateral loads.
Overstrength factor or seismic force amplification factor—a factor applied to structural elements that are expected to remain elastic during the design ground motion to ensure system integrity.
Overturning—the tendency for the structure to overturn due to applied loads.
Peak ground (or maximum) acceleration—the acceleration of a particle on the ground as a result of ground motion.
Perfect tube—a structural system where all members resist either tension or compression and bending is essentially eliminated through the use of a mesh-tube concept. Shear-lag in the tube is essentially eliminated through the densely spaced diagonal members.
Performance-based design—a method of design specifically addresses seismic performance of tall buildings, including structures with long fundamental periods of vibrations, significant mass participation and lateral load-response in higher modes of vibration, and a relatively high aspect ratio (slender profile). The process typically results in a better understanding of structural behavior, but does not lead to enhanced performance unless specific higher performance objectives (minimum objectives are defined in the building code) are used, including ground motion input, components, or systems.
Period—the time it takes for one full oscillation of a structure. The reciprocal of frequency.
Pile cap—structural element used to transfer load from the superstructure to pile systems. Usually comprised of reinforced concrete.
Piles or caissons—vertical or sometimes sloped load-carrying elements used to impose load on suitable soil layers below the surface of the ground. Loads are typically transferred by end bearing and skin friction between the pile or caisson and the soil. Lengths of piles or caissons vary based on applied loads and soil conditions. Piles or caisson lengths can vary from a few feet to several hundred feet.
Pin-Fuse Frame™—a patented seismic frame device that uses friction fuse to dissipate energy during an earthquake—uses pins, circular bolt arrangements for beam end connections, brass shims, long-slotted holes in diagonal braces, and high strength steel bolts.
Pin-Fuse Joint®—a patented seismic frame device that uses friction fuse to dissipate energy during an earthquake—uses a pin, curved steel plates, brass shims, and high strength steel bolts.
Pinned joints—the use of pins to eliminate moment-resistance in a joint.
Pinned-truss concept—the use of pins to allow for movement in a truss system when subjected to relative end displacements.
Pre-consolidated—soils that have deformed over time as a result of load usually associated with clay where the soil "squeezes" over time when subjected to compressive loads.
Pressure grouting—a technique used to strength soil layers. Cement grout is typically injected into the soil to increase strength and limit settlement.

Pressure tap modeling—devices or taps included in the physical model for the tower designed to read actual pressure applied to local building surfaces during wind tunnel testing.

Prestress—placement of loads on a structure to create an initial state of stress usually introduced to offset other applied loads.

Prestressed frames—the use of prestressing to balance load in frames providing strength and deflection control. Typically included in reinforced concrete structures but could be included in steel or composite structures.

Probability of exceedance—the statistical probability of exceedance relative to a specified return period.

Probable Maximum Loss (PML)—the amount of financial loss expected after a prescribed earthquake.

psf—pounds per square foot.

Quadratic formulation—a polynomial equation to the second degree—used in structural engineering mechanics.

Radiated energy—during an earthquake stored energy is transformed and results in cracks/deformations in rock, heat, and radiated energy. Represents only a small fraction of the total amount of energy transformed during an earthquake and is the seismic energy registered on seismographs.

Rational wind tunnel studies—wind studies using modeling techniques that consider actual building properties, site conditions, and historical wind data.

Reduced beam section (RBS)/dogbone connection—a connection that incorporates reduced cross-sectional area in a frame beam just outside of the column face but eliminating part of the flanges.

Redundancy—(a) providing multiple load paths for forces to travel within the structure. If a member or connection loses its strength due to overload, other members and/or connections will participate in resisting the applied forces and reduce potential for localized or overall progressive structural collapse. (b) The duplication of critical structural component to increase the reliability of the overall structural system.

Reinforced concrete—the mixture of cement, sand, aggregate, water, admixtures including accelerators, plasticizers, retarders etc., and reinforcing steel or post-tensioning tendons. Accelerators speed curing (the chemical process of hydration between the cement and water) of the placed mix for conditions such as cold-weather placement of concrete. Retarders slow curing of the placed mix for conditions such as hot-weather placement of concrete.

Relative displacement—unequal displacement between to structural members typically between columns or walls in a floor.

Reliability/redundancy factor—a factor used in the calculation of earthquake forces that considers the reliability I redundancy of a specific structural system.

Resonance—the tendency of the structure to oscillate at larger amplitudes at some periods or frequencies rather than others. When a vibrating structure is displaced at period close to its natural period accelerations may increase significantly as much as four or five times.

Response modification factor—a modification factor that accounts for specific structural system. The factor increases with system ductility. For instance, a steel moment-resisting frame has a higher response modification factor than a steel braced frame.

Response spectrum (spectra)—since seismic ground motion produces transient rather than steady state input the response spectrum is a plot of the peak response (acceleration, velocity or displacement) of a series of oscillators with varying natural frequencies or periods that are forced into motion by the same base vibration or shock. The resulting plot is used to determine the response of a structure with a given natural period of vibration assuming that the system is linear (no permanent deformation of structure under load). Response spectra can be used to assess multiple modes of oscillation (multi-degree of freedom systems that typically include buildings because of multiple locations of mass, i.e. floors), which are only accurate for low levels of damping (most buildings). Modal analysis is performed to identify the modes of vibration and the response for each of those modes can be obtained from the response spectra. Each peak response is then combined for a total response. The typical combination method of these responses is through square root of the sum of the squares (SRSS) provided that the modal frequencies are not close.

Rheological buildings—using structures to accommodate material state changes from liquid to solid, systems that can change their characteristics by altering internal stiffness.

Rigid diaphragms—horizontal-framing systems typically consisting of concrete slabs used to interconnect the structural system together.

Rigid frame—moment-resisting steel, concrete or composite frame that incorporates fully fixed connections.

Screen frames—structural frames comprising structural steel, concrete or composite that are used to resist lateral loads along with shading of the structure or the like—may have infill panels that are not symmetrical.

Seismic base shear—the force applied to the base of a structure due to seismicity. This force is distributed over the height of the structure.

Seismic response coefficient—a coefficient based on spectral response acceleration, a structural system-dependent response modification factor, and an occupancy importance factor. This factor is applied to the weight of the structure to determine the seismic base shear.

Seismic zone factor—a factor associated with mapped zones of anticipated ground accelerations.

Seismometer—instrument used to measure and record motions of the ground.

Self-reflection—designing structures to self-reflect on imposed loads through passive or active systems. Properties of the structure are changed to respond to these loadings.

Self-sufficiency—buildings that are capable of being self-sufficient, not relying on external sources of energy or water. Embedded systems are capable of treating waste. Systems are completely detached from infrastructure grids.

Semi-rigid frame—a moment-resisting steel frame that incorporates connections that has partial fixity allowing for some rotation when loaded.

Sensory fields—a field of sensors including accelerometers and anemometers used to pin point natural force flows on a district or city-wide grid. Advanced analytical models of building structures are interfaced with the sensory field to sense real-time forces and direction of forces.

Service area—typically the area within the core that includes stairs, elevators, mechanical rooms and the like.

Serviceability—performance characteristics of the structure including but not limited to drift, damping, acceleration, creep, shrinkage, and elastic shortening.

Serviceability including drifts and accelerations—considerations for the performance of structures in addition to strength primarily related to imposed lateral loads from wind. Drift is the displacement of the structure (between floors or over the height of the structure) due

to imposed lateral loads. Exterior wall system, building partitions, elevator hoistways etc. must be designed for this displacement. Accelerations result from imposed lateral loads on the structure and must be considered for perception to motion of building occupants.
Settlement and differential settlement—displacement of soil systems when subjected to load. Settlements may be differing across the foundation system due to unsymmetrical loads or varying soil conditions.
Shear keyways—continuous vertical notches placed in slurry walls to transfer shear loads imposed from soil from one panel or segment to the next.
Shear lag—inefficiencies in transferring forces to columns on the faces of the structure perpendicular to the application of load typically in a tubular or mesh tube frame.
Shear load—Load applied across the axis of a member or system.
Shear truss—a steel truss typically consisting of diagonal members and located in the core area of the building designed to resist lateral shear.
Shear wall—composite frame—a reinforced concrete shear wall core combined with a frame that includes columns and beam comprised of both structural steel and reinforced concrete.
Shear wall buttressed core—a structural system that combines a central shear wall core.
Shim—a material used between two other materials that is required to take up dimension or to provide predictable coefficient of friction.
Shrinkage—a phenomenon that occurs in concrete where volume is reduced through the hydration I drying process.
Simplicity—simplicity in form and purity of concept for tall building design where symmetry, uniformity of mass, and control of force flow leads to least use of material and environmental impact.
Skin friction—the friction developed between the surface of a pile or caisson and the neighboring soil.
Slip-critical—typically refers to bolted connections where the threshold of slip is important.
Slurry walls—foundation system typically installed with bentonite slurry. Trenches are typically dug from the surface with bentonite installed to keep neighboring soil from collapsing into the trenches. Bentonite slurry has higher density than neighboring soil to prevent soil from collapsing and is displaced from the trench during the concrete placement process. Walls are typically installed in panels or segments with shear keyways placed in between.
Soil stratum or layers—layers of soil having specific thicknesses and geotechnical structural characteristics.
Spectral acceleration—is approximately the acceleration experienced by the building, as modeled by a particle mass on a massless vertical rod having the same natural period as the building. When the mass-rod system is moved or "pushed" at its base using a seismic record assuming a certain damping to the mass-rod system (typically 5%) one obtains a record of particle motion.
Spectral displacement—is approximately the displacement experienced by the building as modeled by a particle mass on a massless vertical rod having the same natural period as the building. Based on the seismic record, the maximum displacement can be recorded.
Spectral velocity—the rate of change (derivative) of the displacement record with respect to time.

Spread or continuous footings—foundation system that distributes load from the superstructure to the soil system. Usually comprised of reinforced concrete with spread footings square or rectangular in shape and continuous footings having a specific width but continuing along under multiple columns or walls. Continuous footings typically used under perimeter foundation walls.

Static—having no motion, objects acted on by forces that are balanced.

Stayed mast—a central core interconnected to perimeter mega-columns or frames with outrigger trusses or walls.

Steel-plated core—structural steel used in core systems as an alternate to a reinforced concrete shear wall system.

Stiff clay (hardpan)—clay having high compressive resistance with little susceptibility to consolidation or creep over time. The City of Chicago is known for having this soil condition with many tall buildings supported by caissons that bear on this soil layer.

Stiffened screen frame—frame with infill panels that are introduced to provide additional stiffness to resist lateral loads.

Stiffness—the product of a material's modulus of elasticity and moment of inertia or axial area.

Stiffness and softness—structural systems that balance stiffness requirements for the tall building primarily due to wind loadings with softness or ductility related to seismic loading.

Straight-shafted or belled caissons—a caisson will have a constant cross-section along its full length or it may have a straight shaft typically along a majority of its length with an enlarged section at its base shaped like a bell.

Strain energy—The external work done on an elastic member in causing it to distort from its unstressed state is transformed into strain energy.

Strength—capacity of a material or structure to resist applied forces.

Structural clarity—structural systems that may be visually understood, but more importantly include clear load paths for buildings subjected to highly uncertain loading conditions such as seismicity.

Subgrade moduli—the relationship of stress and strain in soil. Used to determine soil stiffness and susceptibility to settlement.

Superframe—a frame located at the perimeter of the structure that incorporates three-dimensionally placed members typically including diagonals.

Superimposed load—permanent load that is constant over time. Loads are due to the weight of exterior wall systems, partitions or the like.

Sustainability—tall buildings designed to be self-sufficient, if not regenerative. Buildings designed to be resilient, incorporating systems that enhance performance and extend service life.

Sustainable Form Inclusion System (SFIS)™—a form inclusion system that introduces postconsumer waste products into structures to reduce waste while reducing structural materials through a reduction in mass.

Tube-in-tube—the combination of frames typically in the center core area and at the perimeter of the structure—the frames typically consist of closely spaced columns having a plan spacing typically similar to the floor-to-floor height.

Unreinforced masonry walls—walls consisting of brick, concrete block or similar materials that do not contain reinforcing steel. May contain cement grout and or mortar.

Uplift—net upward force on a structural element or system—will cause upward displacement if not restrained.

Vertical force distribution—distribution of lateral forces typically due to wind or seismic loadings over the height of the structure. Forces can be applied as distributed loads or point loads at floor diaphragms.
Vortex shedding—the unsteady air flow that takes place in special flow velocities according to the size and shape of the structure. Vortices are created at the back of the structure and detach periodically from either side of the structure. The fluid flow past the object creates alternating low-pressure vortices and the structure tends to move toward those areas of low pressure.
Well-defined and understandable load paths—defining structures where a clear load path is defined by one member supporting another due to gravity or lateral loads. A lateral load resisting system that resists lateral loads through a clearly understood load path from superstructure to foundations.
Wide-flanged beam analogy—force distribution analogy comparing a wide-flanged beam to columns located in the plan of a tall building.
Wind-induced motion—motion induced into a structure by wind forces.
Wind velocity—the speed at which wind travels relative to an at-rest condition.
X-braced frames—frames that introduce crossing diagonals into the truss system sharing work points—may be used on core, perimeter, belt or outrigger truss systems among others.
Zero damping—a system that is absent of any effect that tends to reduce the amplitude of oscillations in an oscillatory system.

REFERENCES

ACI 318-08 *Building Code Requirements for Structural Concrete* American Concrete Institute, Farmington Hills, MI, 2008.

Ali, M. M. *The Art of the Skyscraper: The Genius of Fazlur Khan.* Rizzoli International Publications, Inc., 2001.

Ambrose, J. and Vergun, D. *Simplified Building Design for Wind and Earthquake Forces*, 2nd edition, New York, Wiley, 1990.

American Society of Civil Engineers (ASCE 7-10) *Minimum Design Loads for Buildings and Other Structures*, 2010.

American Society of Civil Engineers (ASCE 88) [formally American National Standards Institute (ANSI 58.1)] *Minimum Design Loads for Buildings and Other Structures*, 1988.

Chinese Building Code, *CECS 230:2008 Specification for Design of Steel-Concrete Mixed Structure of Tall Buildings*, 2008.

Chinese Building Code, *GB 50011-2001 Code for Seismic Design of Buildings*, 2001.

Chinese Building Code, *JGJ 3-2002 Technical Specification for Concrete Structures of Tall Buildings*, 2002.

Chinese Building Code, *JGJ 99-98 Technical Specification for Steel Structures of Tall Buildings*, 1998.

Darwin, C. *The Origin of Species by Means of Natural Selection, or the Preservation of Favoured Races in the Struggle for Life.* London, John Murray, 1859.

Fanella, D. and Munshi, J. *Design of Concrete Buildings for Earthquake and Wind Forces.* Portland Cement Association (PCA) According to the 1997 Uniform Building Code, 1998.

Holland, J. *Adaptation in Natural and Artificial Systems.* Ann Arbor, MI, University of Michigan Press, 1975.

Intergovernmental Panel on Climate Change (IPCC), "Climate Change 2007: Synthesis Report," Contribution of Working Groups I, II, and III to the Fourth Assessment Report of the Intergovernmental Panel on Climate Change, Core Writing Team, eds. R. K. Pachauri and A. Reisinger. Geneva, Switzerland, IPCC, 2007, p. 104.

International Code Council, *International Building Code (IBC) 2012: Code and Commentary, Volume 1*, 2012.

International Conference of Building Officials (ICBO), *Uniform Building Code (UBC).* Whittier, CA, Structural Engineering Design Provisions, 1997.

Janssen, J. J. A. *Mechanical Properties of Bamboo.* Springer, New York, 1991.

Khan, F. R. *ENR*, interview with Fazlur Khan. 1972.

Khan, F. R. *Architectural Club of Chicago, Club Journal*, 1982.

Khan, Y. S. *Engineering Architecture – The Vision of Fazlur R. Khan.* New York, Norton, 2004.

Lindeburg, M. and Baradar, M. *Seismic Design of Building Structures 8th Edition.* Belmont, CA, Professional Publications, Inc., 2001.

National Building Code of Canada (NBC), *Structural Commentary Part 4 –Wind Engineering.* 2005.

PEER/ATC-72-1, *Modeling and Acceptance Criteria for Seismic Design and Analysis of Tall Buildings*. Applied Technology Council, Redwood City, CA, 2010.

Perform-3D V5 User Guide, *Nonlinear Analysis and Performance Assessment for 3D Structures*. Computers and Structures, Inc., Berkeley, CA, 2011.

Perform-3D V5, *Components and Elements for Perform-3D and Perform-Collapse*. Computers and Structures, Inc., Berkeley, CA, 2011.

Sarkisian, M. *Structural Seismic Devices*, United States Patent Nos. US 6,681,538; 7,000,304; 7,712,266; 7,647,734; 8,256,173; 8,452,573.

Sev, A., A. Ozgen. "Space Efficiency in High-Rise Office Buildings," *Journal of the Faculty of Architecture*, 2(4) (2009), 69–89.

Tall Building Initiative, *Guidelines for Performance-Based Seismic Design of Tall Buildings*, Version 1, Report No.2010/5. Pacific Earthquake Engineering Research Center, Berkeley, CA, 2010.

US Green Building Council, *Buildings and Climate Change*. USGBC Report, 2004.

Willis, C. (ed.), *Building the Empire State*. New York, Norton, 1998.

ACKNOWLEDGMENTS

THE WORK

The Architects and Engineers of Skidmore, Owings & Merrill LLP

CONTRIBUTORS TO THE BOOK DEVELOPMENT

Structural Engineering

Eric Long
Neville Mathias
Jeffrey Keileh
Danny Bentley
Rupa Garai
Chris Horiuchi

Original Graphics

Lonny Israel
Brad Thomas

Publication Coordination/Images

Pam Raymond
Harriet Tzou
Rae Quigley
Justina Szal
Brian Pobuda
Brenda Munson
Nika Wynnyk

SOM Integrated Design Studio – Stanford University

Brian Lee
Leo Chow
Mark Sarkisian
Brian Mulder
Eric Long

Editing

Cathy Sarkisian

Inspiration

Stan Korista

Special acknowledgment to Eric Long for his tireless efforts in helping make this book possible.

INDEX

Page numbers in italics indicate figures